A Family Guide To Keeping Chickens

2nd Edition

Anne Perdeaux

....

A How To Book

ROBINSON

ROBINSON

First published in Great Britain in 2014 by
How To Books, an imprint of Constable &
Robinson Ltd.

This edition published in 2018 by Robinson

Text copyright © Anne Perdeaux, 2018

Illustrations © Jess Goodman 2018

10 9 8 7 6 5 4 3 2 1

A CIP catalogue record for this book
is available from the British Library.

ISBN: 978-1-47214-044-9

Typeset in Great Britain by Mousemat
Design Limited

Printed and bound in Great Britain by CPI
Group (UK) Ltd, Croydon CR0 4YY

Papers used by Robinson are from well-
managed forests and other responsible
sources

MIX
Paper from
responsible sources
FSC® C104740

Robinson
An imprint of
Little, Brown Book Group
Carmelite House
50 Victoria Embankment
London EC4Y 0DZ

An Hachette UK Company
www.hachette.co.uk

www.littlebrown.co.uk

How To Books are published by
Robinson, an imprint of Little, Brown
Book Group. We welcome proposals
from authors who have first-hand
experience of their subjects. Please set
out the aims of your book, its target
market and its suggested contents in an
email to Nikki.Read@howtobooks.co.uk

Anne Perdeaux is an expert on chickens. She is a regular contributor to the popular *Your Chickens* magazine, providing articles, quizzes and monthly material for the children's pages. *Country Smallholding* magazine frequently publishes her poultry articles, and for both these magazines she is on a panel of experts answering readers' queries.

To my father, who has always encouraged my enthusiasm for animals, and to Martin, who lives with the consequences

Contents

Preface

Chickens are extremely addictive and, as with many addictions, it's a gentle yet slippery slope. 'Only a couple, just for fun,' you think to yourself. 'What harm can it do?' Then, before you can say 'Cock-a-doodle-do', you are clucking to them, gossiping about them and waking up in a cold sweat worrying about foxes. Soon you will want more and more, realizing too late that you are now one of the millions of chicken addicts – there is no cure.

Well, you have been warned, but if you are still determined, then this book will help you to start keeping chickens. We will explore their strange and wonderful world, finding out what they need to make them happy – and happy hens lay eggs (sometimes).

Why Keep Chickens?

It's not surprising that so many people keep chickens. These versatile creatures can be at home on a farm or in a small back garden. They make excellent family pets, requiring less attention than a dog, while being entertaining, productive and educational. For those seeking the 'good life', they are the obvious first step into keeping livestock.

For hundreds of years, chickens have been valued all over the world – for their eggs, meat, fighting skills, or simply their beauty. The chicken is now enjoying a renaissance, and is also considered a desirable pet, competing with dogs, cats, rabbits and guinea pigs.

For those who have never kept chickens, this can be surprising. Chickens are often regarded as dull, brown, egg-producing birds, or viewed neatly wrapped in the chiller cabinet. Useful, yes, but not necessarily what you would look for in a companion.

Yet chicken keepers quickly discover that there is more going on under all those feathers than you'd think. Chickens have complex social structures, are surprisingly intelligent and quickly demonstrate individual characters.

Although chickens have many endearing traits, most people start keeping them for their eggs. It's very satisfying to be even a little self-sufficient; to stride past the egg shelves in the supermarket, knowing you have better and fresher ones at home.

If you grow vegetables, the addition of eggs from your hens can turn a simple meal into a feast – how about a salad from the garden to accompany a golden omelette, flecked with fresh herbs?

Chicken manure is also great on the vegetable garden, chickens can turn over the soil when a patch needs to be cleared, and you can feed the chickens some of your surplus vegetables. It's a satisfying circle.

Maybe you are now itching to buy your first chickens, but before racing off to 'Chickens-R-Us', please read on a little further. Some careful planning will make the whole experience more enjoyable – for you and, most importantly, for your new chickens.

CHAPTER ONE

Before Your Chickens Come Home to Roost: Planning for Your First Chickens

It's easy to buy a coop and chickens – and it's also easy to make costly mistakes. Don't hurry the process of acquiring your first chickens. Use this book as a guide and take some time to investigate all that will be required.

Can You Keep Chickens?

Rules and regulations

Check there are no restrictions on keeping chickens, either in your house deeds, rental agreement or local by-laws. Even rural areas sometimes have regulations of these kinds. Chickens are usually considered as livestock, including those kept as pets. If in doubt, contact your local authority.

At present you only need to register with DEFRA (Department for Environment, Food and Rural Affairs) if you keep more than fifty birds, including chickens, ducks, turkeys, guinea fowl, etc.

You might wish to speak to your neighbours about your intention to keep chickens. Some people love the idea – and some don't. It's better to calm any concerns they have at this stage rather than wait for a complaint.

Ten Chicken Myths

1. Hens won't lay eggs without a cockerel
Unlike many birds, hens will lay eggs without a male being present.
2. Chickens attract mice and rats
Mice and rats are attracted by chicken feed – this should be cleared away at night and stored securely. Chickens will kill (and eat) mice!
3. Chickens smell

Given the correct conditions chickens produce little odour. If droppings are allowed to build up, they will smell very unpleasant and also affect the chickens' health.

4. Chickens are noisy

Hens aren't silent, but usually confine themselves to gentle clucking unless startled. Some can make a racket when laying an egg, causing the other hens to start a chorus of encouragement, but unless kept in large numbers hens should cause little disturbance. Cockerels can be very noisy.

5. Chickens can be used to clear up household scraps

Feeding kitchen waste to chickens is now illegal in the UK.

6. Chickens spread disease

While there are a number of chicken diseases, very few of these can be transmitted to humans. It's always sensible to wash your hands after being in contact with any animal, and to change clothing if necessary.

7. Chickens must be let out at dawn every morning

Chickens need to be let out, but not necessarily at first light – at dawn the night-time predators will still be about. However, too much restriction of daylight can reduce egg production.

8. Chickens are vegetarians

Chickens are omnivores and are attracted to meat. They will forage for worms and insects as well as vegetable matter.

9. Eggs are either brown or white

Eggs vary from pure white to dark chocolate in colour, even blue or green – it depends on the breed of hen. The shell colour makes no difference to their nutritional value.

10. Eggs contain baby chickens

Infertile eggs will never hatch, and even fertile eggs won't develop unless incubated.

Will chickens fit into your life?

Take a moment to reflect on your circumstances. Chickens are less trouble than many animals, but they still require care and will take up your time.

As a bare minimum, you will need to ensure the chickens always have water and food. They must be safely shut away at night and let out in the morning. Their housing will need cleaning regularly, and their health should be frequently monitored.

That is the minimum, but there will be other jobs too. There will also be times when you have to drop everything and deal with something urgent and chicken-related.

Responsible chicken keeping

Once you establish a routine, caring for chickens is not difficult or laborious. They are fascinating creatures, and you may actually find it quite difficult to tear yourself away from them.

Still, you will probably want to take an occasional break, so before buying your chickens consider who will look after them when you go on holiday.

As with any animals, chickens must be treated responsibly. Apart from the moral issue, there may be complaints to welfare organizations if they are neglected.

The Five Freedoms

The Animal Welfare Act (2006) includes a duty of care for animals. All keepers must take reasonable steps to ensure they meet their animals' welfare needs. These are the 'Five Freedoms' recommended by the Farm Animal Welfare Council:

- Freedom from hunger and thirst
- Freedom from discomfort
- Freedom from pain, injury or disease
- Freedom to express normal behaviour
- Freedom from fear and distress

How much will it cost?

Will chicken-keeping pay off? Will you be able to recycle household scraps, turn them into eggs and become an egg entrepreneur?

It's not easy to make chickens pay for themselves. Your reward is more likely to be pride in your hens, the enjoyment of their company and the benefit of delicious eggs.

Chickens fed an unbalanced diet of scraps won't produce many eggs, and in any case feeding kitchen waste to chickens has been illegal in Britain since the foot-and-mouth disease outbreak of 2001. You will need to buy feed for your hens and, taking the initial expenses into account, your eggs are unlikely to be particularly cheap.

Even so, chickens aren't as costly as many animals to keep, and the eggs are definitely a bonus.

If cost-effective chickens are your aim, you will need to manage them carefully, choosing only productive breeds, and being prepared to take a firm line with any hens that aren't pulling their weight in the egg-laying department.

Finding Out about Chickens

This book will set you on your way, but practical research is indispensable.

Try to make friends with a neighbouring poultry keeper – most chicken enthusiasts love chatting about their hobby. Perhaps you could offer to look after the chickens while the owner is away. This will provide useful experience, and could even be the start of a 'hen-sitting circle'.

It's worth taking plenty of time to look at different types of chicken housing, comparing prices with quality and design (see Chapter 3). Country shows usually have poultry trade-stands as well as exhibitions of chickens, offering an excellent opportunity to browse and quiz the experts. Many animal feed stores also sell poultry equipment, and staff are often able to provide advice. Chicken breeders, poultry shows and auctions are also well worth a visit.

Being prepared

Apparently people sometimes buy chickens on impulse, without having anywhere to keep them. I wouldn't recommend this – it's very stressful.

Due to a series of perfectly avoidable delays, our chicken house was erected only a couple of hours before the chickens were due – darkness was falling as the last screws were tightened. There was no time to waste, as we had bought the flock from a neighbour who was moving the next day. Twelve chickens were transferred into their new home in pitch blackness as the heavens opened. (Incidentally, I had originally planned to take only four, which shows how dangerously addictive chickens can be – even before they arrive.)

Chickens as Pets

Chickens can make good pets, becoming very tame and affectionate towards their owners while retaining a measure of independence. That is, they require care and enjoy human company, but won't pine if you leave them for a few hours. They'll rush to meet you when you arrive home exhausted, and provide a soothing break from the day's stresses and strains.

Watching chickens is very therapeutic – and they lay eggs too, which is more than can be said for most therapists.

Chickens and children

Children are usually fascinated by chickens and they are ideal children's pets, giving an object lesson in where food comes from, as well as being companionable and fun at the same time.

Very young children enjoy watching chickens and helping to collect the eggs.

Small fingers thrusting through the wire might receive an inquisitive peck, but nothing like the nip inflicted by the sharp teeth of a rabbit or guinea pig. Older children can easily learn to look after chickens, gaining a sense of achievement in supplying the family with eggs or selling the surplus to increase their pocket money.

Nothing New About Keeping Pet Chickens

In the American Civil War, Confederate General Robert E. Lee had a hen called Nellie who laid an egg under his bunk every morning. She went missing after the Battle of Gettysburg, and the retreat was delayed while everyone searched for her. Only when she was found nesting happily in an ambulance wagon could the defeated soldiers pack up and go.

In the same war, the 3rd Tennessee Infantry Regiment had a pet rooster called Jake. During the Battle of Fort Donelson, the rooster encouraged the troops by screeching defiance at the enemy, and accompanied them into prison when the fort was captured. Seven months later the regiment returned to Tennessee, where Jake received a hero's welcome. He died soon afterwards and was given a military funeral. A portrait of him has now been restored and includes a brass plaque summarizing his military career.

There was also a chicken who learned to parachute – we'll come to her later . . .

Key Points

- Check there are no restrictions on keeping chickens at your home
- Registration with DEFRA is voluntary unless you keep more than fifty birds
- Hens don't need a cockerel to lay eggs
- Infertile eggs can't hatch and fertile eggs only develop if incubated
- Good management will prevent smells and other problems
- Feeding kitchen waste to chickens is now illegal
- Chickens are omnivores
- Make sure you have time to look after chickens – you will be legally responsible for their welfare
- Consider what will happen when you are away from home
- Keeping chickens is fun – but they are unlikely to pay for themselves in eggs
- Find out as much as possible before buying any equipment or chickens
- Chickens make good pets and are fairly independent

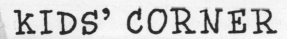

KIDS' CORNER

Quiz

How much have you found out about chickens already?
Test yourself!

QUESTION ONE
What do chickens eat?
- (a) Nothing but meat
- (b) Only vegetables
- (c) Both meat and vegetable products

QUESTION TWO
Which of these statements is true?
- (a) You don't need to keep a cockerel for hens to lay eggs
- (b) Chicken eggs are only brown or white
- (c) All eggs have a baby chick inside

QUESTION THREE
How easy is it to keep chickens?
- (a) They can be left to look after themselves
- (b) Chickens need daily care
- (c) Only farmers know how to keep chickens

QUESTION FOUR
You've decided you want to keep chickens – what should you do first?
- (a) Go out and buy some chickens
- (b) Order a chicken house
- (c) Find out as much as you can before buying anything

QUESTION FIVE
Do chickens make good pets?
- (a) No, they are very fierce
- (b) Yes, they are usually friendly birds
- (c) They aren't safe around children

ANSWERS
One (c); Two (a); Three (b); Four (c); Five (b)

Did you get all the right answers? Well done if you did! Have another look at Chapter 1 if you're not sure about anything.

Chicken Chat

How many times do we talk about chickens without even realizing it? What about: 'He's a real chicken!'? This is often used to describe someone cowardly, because chickens tend to run away from danger. It's rather unfair though, as some chickens can be very brave. In this chapter you read about Jake, the heroic rooster who fought in the American Civil War, and in Chapter 8 there's the story of Myrtle, the skydiving chicken. Both chickens – but neither of them was 'chicken'!

Chicken Jokes

There are an amazing amount of chicken jokes around – here's an old favourite, with a twist:

Why did the chicken run across the road?
Because there was a car coming!

Who tells the best chicken jokes?
Comedi-hens!

Something to do . . .

Do any of your friends keep chickens? Ask whether they enjoy having chickens at home and how they look after them.

CHAPTER TWO

What Really Goes On at a Hen Party? Understanding Chicken Behaviour

Understanding chickens will help you decide how best to keep them – whether you have a small back garden or several acres.

The Intelligent Chicken

It's becoming increasingly apparent that chickens are more intelligent than previously supposed – they can even be trained – and a new generation of chicken lovers are discovering an intriguing, often underestimated creature.

Although you won't find chickens scratching complex mathematical equations in the dust, it soon becomes obvious that a flock consists of a variety of individuals living in an organized society. They have an early-warning system to alert them to danger, communicate in a variety of different tones and display a structure to their daily lives. We have lost so much of our own instinctive behaviour that watching chickens can be a humbling experience.

A flock isn't immediately created when a group of chickens are put together. First they must sort out their hierarchy.

Establishing the Pecking Order

We even use the phrase 'pecking order' about ourselves, as in, 'Now he's had a pay rise, he's definitely top of the pecking order'.

This organization is fundamental to the social structure and safety of the flock. There will be a clear line of succession.

First comes the chief hen (although if you have a cockerel, he will automatically take the principal role – chickens don't do political correctness). The chief hen will be first to eat and drink, seeing off any challenges with a swift, sharp peck. She will sleep on the highest perch, take precedence in the nest-box and enjoy the best dust-bath. With power comes responsibility, and she will warn the others of danger, even going into battle on their behalf.

There will be a second hen and a third, and so on. The hen at the bottom of the pecking order will be last to eat, and roost on the lowest perch. If she steps out of line, there will be several senior hens to put her in her place! Some chickens settle their pecking order without too much trouble, but there can be some spectacular fights in a new flock before the leaders are established.

Understanding the significance of the pecking order helps in managing chickens, as well as making decisions about housing and breeds.

Changes to the pecking order

Once the pecking order has been resolved, peace will reign until the set-up of the flock is changed. If a bird has to be removed temporarily, she may lose her place and have to regain it. A sick bird will be picked on and may be expelled from the flock, which is why chickens don't show ill-health easily.

Those lower down the pecking order sometimes grasp an opportunity to move up a place – perhaps if a higher bird is off-colour. This entails a challenge, which may or may not result in a change to the set-up.

Introducing new birds to an existing flock will precipitate a reshuffle with the accompanying fights for dominance.

Did You Know?

DNA tests have discovered chickens are the closest living descendants of Tyrannosaurus Rex!

Once the chickens have sorted out their hierarchy, they will settle down to their favourite activity.

Scratching and Foraging

Put chickens in a wooded area and you will immediately see their jungle-fowl ancestry. Turning over leaves to search for insects and grubs keeps them happily occupied all day. If a wood isn't available, they will easily adapt to flowerbeds, gravel or lawn. They love to eat grass too – access to grass will provide eggs with the prized deep-yellow yolks.

Some breeds are particularly enthusiastic foragers, and given enough space will almost feed themselves. These birds are great for free-ranging, but preferably not in immaculate gardens.

If grass is in short supply chickens will take whatever green stuff is available. A small garden can quickly look as if a plague of biblical proportions has passed through. This doesn't mean you have to sacrifice your garden to keep chickens happy. They can be contained in a static or movable run, but will need some diversions to keep them from becoming bored.

When not eating, chickens need to spend time on their beauty routines – preening and dust-bathing are vital to their well-being.

The Importance of Dust-Bathing

While garden birds enjoy splashing in a bird-bath, chickens use dust to cleanse themselves and help control parasites. After a good long soak in a dust-bath, they spend hours preening their feathers until they are ready to again join the hen party.

If your chickens are confined to a run, you should provide a dust-bath (see Chapter 5). Free-range birds will find a patch of dry soil and make their own bath – it's a remarkable process to watch.

The chicken starts by loosening soil with its beak and feet. A chicken-sized dip appears, becoming deeper as the bird wriggles down into it. Clouds of dust fly up as the chicken kicks its legs and shakes soil through its wings. After such a burst of activity, there's a blissful rest (you'll feel like providing a book and candles), with just the occasional shuffle to tip more dust into the feathers.

Several chickens may make dust-baths next to each other, or they might make one big bath, like a feathery rugby team.

In the garden, these home improvements may not find favour with the head gardener. Supplying the chickens with a dust-bath can help avoid damage to the flowerbeds or lawn – although they may ignore it and do their own thing.

After bathing the chicken uses its beak to tidy and align each feather, applying preen oil from a gland at the base of the tail.

Now they have had a bath, eaten everything in sight and sorted out their social life, it's time for the chickens to fulfil their side of the bargain.

Laying Eggs

This useful trait has been exploited by humans for centuries, using selective breeding to increase egg production and extend the laying season.

All hens lay eggs, but in varying quantities, according to their breed. Some only lay in spring and summer, while commercial layers are productive throughout the year.

Battery hens were developed as egg producers and are kept in restricted conditions to ensure all their energy goes into laying eggs. Having just considered some of the chicken's basic needs, you may be wondering about the very different existence of a battery hen.

Battery Hens

In January 2012 European legislation outlawed cramped battery cages, in which most commercial laying hens lived out their short existence. These cages allowed very little freedom of movement and no possibility of expressing natural behaviour.

In the UK, at least, farmers have complied with the regulations and changed to 'enriched cages'. Although still not ideal, these do allow the hens some potential to perch, scratch, nest and stretch.

The British Hen Welfare Trust (BHWT) encourages consumers to support British farmers, raising awareness of products made from imported eggs, which may have been laid by hens still kept in the old-style battery cages. Their ultimate aim is for all hens to have outdoor access. The BHWT also re-homes commercial hens that would otherwise be sent for slaughter when egg production declines after their first year of intensive laying (see Chapter 4).

In a domestic situation your chickens can enjoy a good quality of life and provide eggs as well. You will quickly notice the difference between eggs laid by your own contented hens and those produced commercially.

See Chapter 10 for more information on egg-laying.

Broody hens

Some hens are very keen to raise the next generation and will 'go broody' (see Chapter 13). This is useful if you want to hatch some chicks, but broody hens don't lay eggs. Not all varieties of chickens go broody, so if eggs are important this should be considered when choosing a breed.

Egg-laying is hard work, and chickens need their rest, so:

What Happens at Bedtime?

Those new to chickens sometimes worry that they will need rounding up at night, but once they know where their home is the chickens will troop into the henhouse and settle down like a group of unruly children. There might be a few arguments over who sits where, but then they will go to sleep. If you check on them with a torch, some may come to life and realize they don't like their neighbours: 'Ugh, I didn't know it was you!'– peck-peck-peck.

The dark makes chickens sleepy – like throwing a cloth over a budgie's cage – and they will be easy to pick up and handle. This is a good time to do anything that requires amenable chickens.

Although it may not look very comfortable, it's better for chickens to roost on perches, and they usually do so.

In the morning the chickens will emerge from their house – often in an order of precedence.

Providing Shade and Shelter

Chickens don't cope well with heat, and are more likely to be distressed from overheating than from cold. If possible they will spend hot days under bushes and hedges – otherwise they should be provided with shade.

Even so, chickens also enjoy sunbathing. They will stretch out, wings spread, looking rather peculiar. As long as they can retreat to the shade when necessary, there is no cause for concern.

Chickens often don't return to their house during the daytime, even in bad weather, so they should have some outdoor shelter or a windbreak to protect them from the elements.

How Do Chickens Get On with Humans?

Many breeds are easy to tame and quickly become part of the family – some pet hens will happily sit on their owners' laps.

There are also breeds that are shy or rather aloof. These will need more coaxing to make friends and might never enjoy being handled.

Hens are unlikely to attack humans, although interfering with a broody hen may result in an indignant peck, and a small child scattering corn could be unnerved by the enthusiastic stampede of a hungry flock.

Cockerels are a different story: some can be very aggressive. Generally a cockerel is likely to be more trouble than he is worth for the beginner (see Chapter 12).

Chickens and Other Animals

Make sure your chickens are not exposed to stress or danger from household and neighbourhood pets. Dogs pose a threat to poultry, although cats may be quickly put in their place – a friend found her hens scoffing the cat food while four cats watched in horror! Cats will take chicks, though, and very small chickens could also be at risk.

There are many chicken predators, but it's a wild world and chickens are also fond of meat. Pet rabbits and guinea pigs should be kept separately from chickens, who may have a go at them, and tiny pets such as mice or reptiles won't last long in the company of greedy chickens.

Farm animals may accidentally step on or kick chickens, but are unlikely to attack them, even when they help themselves from their feed buckets – although pigs have been known to eat unwary hens who venture into their pen.

Getting to know your chickens

Even when kept mainly for eggs, chickens can provide pleasure and interest for their owners. By getting to know your chickens you will quickly learn to recognize any odd behaviour or problems.

Whether as pets or livestock, keeping chickens is fun and rewarding. They will repay your care and attention in many ways, as well as providing delicious eggs. It's not unusual for people to start keeping chickens for their eggs, only to find that they have also acquired some delightful pets.

Key Points

- Consider chickens' basic needs when deciding how you will keep them
- Establishing a pecking order is essential, but can result in fights
- Scratching and foraging is instinctive natural behaviour
- Dust-bathing is essential for healthy, happy chickens
- All hens lay eggs, but quantities vary according to breed
- Chickens naturally go to roost at dusk
- Darkness makes chickens sleepy and easy to handle
- Chickens need shade in hot weather and shelter from the elements
- Hens are usually docile with humans, but cockerels can be aggressive
- Chickens can be at risk from some pets, but may also attack small animals
- Getting to know your chickens helps you to spot problems early

KIDS' CORNER

Quiz

Chickens have some funny habits – what did you learn from Chapter 2?

QUESTION ONE

What is the pecking order?
- (a) The order in which a chicken eats its food
- (b) A command that makes chickens peck people
- (c) The way chickens organize their flock

QUESTION TWO

Chickens are the closest living descendants of which dinosaur?
- (a) Tyrannosaurus Rex
- (b) Brontosaurus
- (c) Pterodactyl

QUESTION THREE

How do chickens keep themselves clean?
- (a) By splashing in ponds
- (b) By taking dust-baths
- (c) They have to be washed by their owners

QUESTION FOUR

What do chickens do when it gets dark?
- (a) Go into their house
- (b) Come out of their house
- (c) Carry on playing until they are put to bed

QUESTION FIVE

Which of these pets are most likely to attack chickens?
- (a) Guinea pigs
- (b) Cats
- (c) Dogs

ANSWERS

One (c); Two (a); Three (b); Four (a); Five (c)

How did you do? Look at Chapter 2 again if you need to know more about chicken behaviour.

Chicken Chat

'Ruling the roost': you have just read about the pecking order, and now you know there is always a chief chicken in the henhouse (or roost). This saying is also used about a person who takes charge of a group of others. For example: 'In our class Paul rules the roost – he always chooses which games we play'.

Chicken Jokes

Why did the chicken run onto the football pitch?
He heard the ref call 'Foul!'

Why did the chicken work hard at school?
So she would do well in her eggs-ams!

Something to do . . .

Try to find some chickens to watch. Stay still and quiet. What are the chickens doing? Can you spot the chief chicken? Are any laying eggs, taking dust-baths or scratching for food? Let the chickens teach you how they like to live!

CHAPTER THREE

Feathering the Nest:
Where and How Will Your Chickens Live?

Whether you have a small garden or a country estate, chickens can become part of your life!

Keeping Chickens in a Small Garden

Gardens are getting smaller, while outdoor living becomes more popular. The garden may already be accommodating toys, a barbecue, even some flowers – and chickens may seem an impossible dream.

To avoid starting a factory farm in the garden, try to allow at least two square metres per chicken. You will also need room for a henhouse, unless it is raised so the chickens can use the area underneath.

If you have a little more space, don't be tempted to cram in extra chickens. It would be better to keep three hens and give them as much room as possible.

Even three chickens will quickly use their scratching and foraging inclinations to turn a small area into a smelly mud-bath – not something you want to look at when sipping a glass of wine on the patio. One solution is to place the run on concrete and provide plenty of scratching material. If droppings are regularly picked up and a sanitizing product is used, odours will be kept to a minimum. The entire contents of the run can be regularly cleared away and the base hosed down.

Using a solid base also prevents vermin and predators from tunnelling in, but if the run is situated directly on the ground the chickens will benefit from any insects they can dig up. The soil should be protected by a thick layer of scratching material, which may get dug into the ground, especially if it becomes wet. Regular picking out of droppings, raking and disinfecting will help. Whenever the run is thoroughly cleaned out, the soil underneath should be treated with a ground sanitizer.

If you have to deal with a run that has already become sodden, you will need to dig out the wet soil first; otherwise whatever you add will quickly become incorporated into the mire.

Ground cover for chicken runs

The success of whichever material you choose depends on careful planning and your own commitment. Try to situate the run in a sheltered place, avoiding low-lying areas. Roofing all or part of the run will help – as long as the chickens still have access to sunshine.

A quick daily clean prevents a build-up of droppings and disease. Neglected chicken runs are bad for the birds and unpleasant for everyone in the vicinity.

Hardwood chips

These should last for a few months with careful maintenance. They are available from poultry stores and DIY shops, or you may be able to obtain them more cheaply from a tree-surgeon (but be careful they don't contain any poisonous yew). Don't use shredded bark as this contains harmful spores.

Rubber chippings

Make sure you buy the type suitable for chickens. They are laid over a weed-proof membrane and should be regularly hosed down. You must allow drainage for the mucky water to escape, or a residue of droppings will become trapped by the membrane.

Straw

Sometimes used in covered runs, straw needs changing regularly, especially if it gets wet, and may cause digestive problems if the chickens eat it. Don't use hay as it contains harmful moulds and quickly turns into a matted mess.

Dust-extracted wood shavings

These are mostly used for bedding, but shavings can be spread thickly in a covered run. Like straw, they need raking and changing frequently, and may get blown around the garden.

Shredded hemp bedding

Originally sold for horses, this is increasingly used for chickens and can be bought from poultry suppliers as well as country stores. It is more expensive than wood shavings, but also more absorbent and composts faster. Although mainly used for bedding, it can also be used in covered runs.

With care, chickens can become an attractive feature of even a small garden. It will take a little effort to keep their surroundings clean and fragrant, but you will have an ever-changing living tableau to enjoy – who needs a fountain?

Keeping Chickens in a Medium-Sized Garden

With more space, your options increase. You could use movable poultry housing to give the chickens access to grass, or free-ranging may be possible.

A large run at the bottom of the garden is the conventional way of keeping chickens, but this quickly becomes a sea of mud or dust. Apart from looking unattractive, the run will also be a breeding-ground for parasites and disease. You could maintain it as suggested for a small run, but this will be both labour-intensive and costly in a larger area.

A better idea is to divide the run in half. The ground can then be treated and rested in one half, while the chickens enjoy fresh grass in the other. Some houses have pop-holes (the little doors the chickens use) on different sides, making this system easy to operate.

Erecting predator-proof fencing around a large run will require some effort and expense.

Predator-proof chicken runs

There are many chicken predators, although foxes are the most feared. They are a formidable foe, being able to climb like cats, jump, dig and bite through wire or wood. They are also patient and resourceful. Foxes are frequently seen in urban gardens and may strike during the day as well as at night.

It's almost impossible to guarantee that any fence will be totally predator-proof, and much depends on how well it is maintained. Constant checking for any weak spots, damage or signs of digging is essential. Tunnelling rabbits can be a problem if their excavations provide an easy route for the fox.

Installing a secure chicken run is a major project, and expensive too. Inferior materials won't provide adequate protection – a reliable supplier should be able to advise on what is required.

Chicken wire isn't proof against much more than chickens and won't give a fox any trouble. Strong welded mesh is required (the lower the gauge, the stronger it will be – choose twelve- or fourteen-gauge for maximum security).

Predators can gain access through surprisingly small spaces – and if the chickens can stick their heads out, they may get them bitten off! Small rectangular mesh is more difficult to bite into, and if it's fine enough should keep out rodents too.

The supports for the wire should be solid timber or metal (timber will deteriorate over time) and sturdy fixings are essential. The door is often the weakest point, so pay particular attention to the construction and fit.

A wire roof will keep out both predators and wild birds. Otherwise the run should be more than 2 metres high, with an overhang of about 50cm angled outwards.

Sink the fence a good 50cm into the ground and then turn it outwards another 50cm. Another option is to surround the run with concrete slabs or mesh laid flat to the ground and pegged down securely. Don't forget the area under the door – using paving to prevent digging will also keep the doorway from becoming muddy.

For extra security a strand of electrified wire can be added around the bottom of the run – and the top as well, if necessary.

However good the fencing, chickens should always be securely shut into their house at night.

Movable chicken housing

Using a movable house and run allows the chickens to enjoy grass without destroying it. They will also be happy to clear your vegetable patch in winter – just move their run to the required area and leave them to it. The droppings are great fertilizer, but they quickly dry and disappear so keeping chickens on the lawn need not be a problem.

Regularly moving the chickens to fresh ground prevents droppings and disease from accumulating. The smaller the run, the more often it will need moving to a place with fresh grass.

This frequent movement also helps to keep the chickens' boundaries secure. In a fixed run their scratching will loosen the soil, allowing easy access to rodents and predators. Given time the chickens may even manage to dig themselves out.

Predators will find it more difficult to tunnel into a run that is constantly moved to firm and level new ground. Fixing wire underneath the run increases security, but also stops the chickens scratching – denying them their natural behaviour.

Free-ranging

Chickens enjoy free-ranging outside

A movable house and run

their run where possible, but giving them total free-range in an average garden may not be practical. It's easier (and kinder) to start off by allowing the hens out occasionally, rather than giving them unrestricted free-range, discovering it doesn't work and then having to confine them.

Free-Ranging in the Garden

Having free-range hens in the garden is delightful. They gather under the table when you eat outdoors and help enthusiastically with the weeding.

It's a great life for the chickens too, but they will give the garden some abuse, so unrestricted free-range should be considered carefully.

Their scratching and foraging doesn't promote a tidy garden – even a large area will suffer some damage. Digging up seedlings, kicking stones onto the lawn to wreck the mower, making dust-baths and eating plants are amongst the chickens' less desirable habits. Even if your own garden can cope, your chickens may find their way into adjoining plots and upset your neighbours.

Garden fencing must be chicken-proof (clipping their wings may stop them flying but chickens are good at squeezing through gaps) and provide a barrier against neighbouring dogs. A normal garden fence will offer little resistance to foxes, which are often around in the daytime, but surrounding your garden with fox-proof fencing is unlikely to be practical and may add a rather prison-like air to your property.

Droppings will be left wherever chickens roam – lawn, patio, garden furniture – so you will quickly learn to check shoes before coming indoors.

Sometimes the chickens even follow you inside. They are nosy creatures and like to explore, although you'd think there was plenty for them to do outside. We had a hen who made her way through the house, up the stairs and along the landing, where she surprised my husband who was sitting at the computer. Perhaps she wanted to check her egg-mails. More likely she was searching for a nesting spot – I heard of a hen who liked to lay her eggs in her owner's wardrobe!

Keeping Chickens in a Large Garden or Smallholding

If you have plenty of space, free-range hens may be possible. Instead of fencing them in, you could fence them out of any areas you wish to preserve. If you have a wooded area or an orchard, they will be very happy. The hens should be shut in at night, and you may need to protect them from daytime predators too.

Depending on the number of chickens and the amount of space, it may be necessary to rotate their ranging area. Electric poultry netting keeps the chickens where you want them (usually) and the foxes out (nearly always).

Electric poultry netting

Less expensive than installing a permanent run, electric netting can be easily moved to give chickens access to fresh ground.

It is also effective in keeping out predators. Although a fox could easily jump over it, they rarely seem to try. Foxes are more inclined to go through or climb, and the shock they receive from an electric fence is usually enough to make them try their luck elsewhere.

The fence won't harm the fox, but the animal will respect it – providing it is adequate and properly maintained. A weak fence is no barrier to a hungry fox.

Electric fencing can be powered by mains electricity or a battery. Mains provides a constant power source (unless there's a power cut), while batteries need regular monitoring and charging. A rechargeable 12-volt leisure battery is preferable to a car battery.

Fencing can be supplied as a complete kit, and the supplier should advise on individual needs. A basic kit consists of netting, posts, an earth spike and an energizer. The energizer transforms the power supplied to produce an uncomfortable electric shock and is a major component of the fence. If the energizer is inadequate, the fence won't be effective.

Poultry netting requires a more powerful energizer than tape or wire – partly because netting takes more power but also to ensure a firm impression on a determined predator. Moreover, chickens are well insulated by their feathers, and may squeeze through a weak fence to meet their fate on the other side!

Select an energizer that can run more fencing than required – this will allow for any leakage of power. Various factors cause power to drain from fencing, most commonly vegetation. Keep grass trimmed short and cut back any overhanging shrubs.

The earth spike is another essential part of the system – make sure it is adequate for your soil and only use the proper connecting cables. Very dry, sandy soil may require extra earths.

Netting is usually just over a metre high, although it's possible to buy higher fencing if necessary. Usually supplied in green, it needn't look obtrusive in the garden or field. Only the horizontal wires are electrified (apart from the bottom one). To avoid leakage of power, don't let the electrified wires come in contact with anything not insulated from the ground – such as wooden posts – and use only the proper connectors for joins. Tension the fence carefully so that the lowest electrified strand isn't touching the ground or vegetation. The fence should be checked regularly – a fence tester eliminates the need for heroics!

A 'gate' can be installed in the netting if required, and you should display a warning notice in areas where the public have access.

Correctly installed fencing only produces an unpleasant sting – children and pets will quickly learn not to touch it. However, small animals and amphibians can become entangled in the netting and continue receiving shocks, which may prove fatal (hedgehogs tend to curl into a ball around the fence). Disconnecting the second line of electrified wire can help prevent this.

Electric netting isn't suitable for waterfowl (they try to walk through it) and shouldn't be used with horned animals.

Farmyard Chickens

You may have seen farms where chickens roam freely and are left to find their own roosts in the barns. Although this wouldn't suit all breeds, some chickens were originally bred to live in the farmyard.

If you have the facilities and like this idea, bear in mind that the chickens will be at risk from predators both day and night. Chicks will be especially vulnerable. Egg collecting is time consuming when nests are made anywhere and everywhere – it's difficult to keep track of freshness too. Farmyard chickens still need food and fresh water, as well as daily checks in case of illness or injury.

Avian Influenza Restrictions

Many chicken keepers were caught napping in December 2016 when, due to outbreaks of avian influenza ('bird flu') in Europe, DEFRA ordered that all poultry should be kept under cover to reduce the risk of infection via migratory wild birds. The disease is spread through direct contact and droppings, so poultry either had to be housed in an outbuilding or contained in a run – covered over to stop any droppings from dropping in. Even moveable housing was to remain unmoved, as fresh ground could be a source of contamination.

These restrictions were legally binding, and weren't lifted until at least the end of February – longer in some areas that were deemed to be at higher risk.

It certainly proved an exercise in ingenuity for those with free-range chickens and no suitable outbuildings in which to confine them. Most rose to the challenge and found a way, but it's possible a similar situation will arise again. It's mentioned here as you may wish to take this into consideration when planning your chicken housing.

There is more about avian influenza in Chapter 11.

Considering How Many Chickens You Can Accommodate

However you decide to keep your chickens, it's essential not to overstock. Trying to keep too many chickens in too little space will result in stress, disease and behavioural problems.

When deciding how many chickens to buy, err on the side of caution. Better to start with a few happy healthy birds than lots of miserable ones.

Spacious living

If your chickens are to be wholly or mainly confined to a run, allow as much space per bird as possible. Manufacturers can sometimes be over-ambitious about how many birds can comfortably live in their chicken runs, or make the assumption they will only be shut in for limited periods. Check the actual measurements, and see if the run can be extended if necessary.

The house should allow all the birds to roost comfortably. Chickens are susceptible to overheating, and a crowded henhouse will lead to hot, unhealthy hens. As chickens do most of their droppings at night, you can imagine how unpleasant the air quality will be if the house is tightly packed. This will harm their respiratory systems.

Chickens are better able to cope with cold than with heat, and will keep each other warm as long as their house isn't too big or draughty. If you're likely to increase your flock in the future, buy a house with reasonable space for expansion, but not so vast that the chickens rattle around in it trying to keep warm. If the addiction really takes hold, you may need to buy another house eventually, but by then you'll have plenty of use for two houses anyway!

Positioning the Chicken House

Choose a well-drained spot for the chicken house. Soggy ground will quickly become a disaster zone and muddy feet lead to dirty eggs.

New chicken keepers often put their chicken house proudly in the middle of the lawn, but chickens are descended from jungle fowl and can become anxious in open spaces where they are exposed to predators. Nervous chickens don't lay eggs, so try to give your chickens the security of trees, a hedge or a fence near to their house.

Hot chickens don't lay eggs either, and a house without any shade could become overheated on a scorching summer's day. However, chickens do enjoy some sunshine, so their run needn't be situated in the gloomiest part of the garden either.

Try to position the pop-hole (the chickens' door) away from the prevailing wind and rain. If this is not possible, you could rig up a screen of some kind (perhaps using fencing or straw bales) to give the chickens some protection from the worst of the weather.

Different Types of Chicken Housing

Before the chicken-keeping craze took off, there was little choice when buying a henhouse. A handful of manufacturers produced a range of traditional designs, some of a better standard than others, and that was that. Nowadays chicken owners can choose between luxury, picturesque, quirky, practical and economy homes.

Basically the alternatives are a house with an attached or integrated run or stand-alone housing.

Integrated runs and arks

A 'poultry ark' is typically a triangular structure incorporating the sleeping quarters and the run, although ark is sometimes used to describe any combination of house and run or simply a triangular-shaped henhouse.

The Boughton chicken ark with sleeping quarters above the run

The shape is said to have evolved as a way of stopping sheep from jumping onto the roof, but it also allows rain to run off efficiently. The triangular design limits perching space, though, which usually isn't high enough for a large chicken or cockerel to roost comfortably without damaging its comb.

The sleeping area of an ark may be situated above the run, which makes good use of space and keeps the run dry too. The chickens go up a ramp to roost – it isn't suitable for tiny chicks.

A small chicken ark – suitable for bantams

When considering an ark, look at the access to the run (working out how easy it would be to catch a chicken). Check the door to the sleeping quarters too. Sometimes the whole side of the roof has to be raised. We had an ark like this and it was a work of art to balance the door open enough to see the occupants while not allowing them to escape.

A chicken house with an attached run may be a better option than an ark. There are several different designs, but if space is limited, think high-rise. A house raised above the ground allows the chickens to make use of the area beneath.

This ark has the sleeping quarters at one end, and could be used as a broody coop

Some manufacturers supply bolt-on extensions for their chicken runs and arks, so you can increase space as required.

Arks and houses with integrated runs are sometimes fitted with wheels or skids so they can be easily moved around. Some smaller arks have handles at each end and can be moved by two people – provided you have a willing accomplice.

Houses with wheels can be easily relocated

Stand-alone houses

These range from the small and bijou to large structures capable of housing a hundred or so birds. You can even buy houses designed to look like gypsy caravans or cottages – anything but henhouses. Designs with wheels or skids allow you to relocate the henhouse easily.

Conventional garden structures such as sheds and wooden playhouses can also be converted to chicken housing provided they are structurally sound, secure and weatherproof.

What to Look for When Buying Chicken Housing

Your needs may vary, depending on whether your henhouse will be in a secluded garden or on a blustery mountainside, but good design and workmanship should be a priority.

Structure

There are some cheap and not so cheerful henhouses available, but paying a little extra usually proves worthwhile. Good-quality timber is expensive, and a henhouse has to incorporate a number of features. Look for solid construction and substantial, pressure-treated timber.

Various plastic houses are now available, and are easy to maintain and clean. Make sure that a non-wooden house has good insulation, and in particular that there is plenty of ventilation. Check that it is stable and cannot be easily blown over.

Whichever house you choose, it must be watertight – look for any places where rain could penetrate. The roof should have a good pitch with an overhang that will shed water away from the house. Roofing felt is a haven for red mite (a nasty, blood-sucking parasite – see Chapter 11), so corrugated bitumen or timber is preferable. Corrugated metal roofing can cause condensation unless it is well insulated.

Be careful if buying second-hand housing – it could harbour disease and parasites. Red mite can remain dormant for months in an empty henhouse, only to spring back to life when new chickens arrive. It is very difficult to eradicate, so you really don't want to start chicken-keeping with a house full of red mite.

There's a wide range of housing available from online stores, but examine the actual house before purchasing if possible. You need to judge quality, practicality of design and ease of use, which is difficult to do from a picture. If shopping online, watch out for cheap imitations of reputable housing.

A cut-price, inadequate house won't seem such a bargain when you're lying in bed, listening to a storm and worrying about your chickens. Finding the roof halfway down the garden the next morning is traumatic too – as I know from personal experience!

Security

Keeping chickens inside is fairly easy. Keeping predators out is more challenging.

When assessing a chicken house (and run), imagine a really determined large animal trying to gain access. Could you get in by using brute force?

Foxes and badgers have incredibly strong jaws and teeth – they can bite through flimsy wood, even wire. Check for any weak spots, particularly around the doors and nest-boxes.

Fixings, hinges and bolts should be robust. Bolts are important – foxes have been known to turn swivel catches. You may also need to add padlocks to prevent human predators from helping themselves to eggs or chickens.

Any windows or ventilation holes should be covered in small-gauge welded mesh to keep out rodents.

If the house has a run attached, it should be constructed of strong wire mesh (not chicken wire) and have solid supports. There should also be a roof to prevent predators from climbing in.

A wire 'skirt' can be laid horizontally around the bottom of the run, making it difficult for animals to burrow underneath.

See also the section on 'Predator-proof chicken runs' earlier in this chapter.

Check for robust fixings

Ventilation

This is vital to the health of your chickens. Being shut in an airless coop all night doesn't sound much fun, and poor air quality will cause sickness in the flock. While good airflow is essential in summer, even on the coldest nights the chickens will require adequate ventilation.

Chickens can tolerate cold fairly well, but draughts are a different matter, so ventilation holes shouldn't be positioned opposite each other. Some houses have sliding vents that allow the airflow to be adjusted – check they are easy to open and close.

There may also be a wire-mesh window, with a see-through cover that can be opened to increase ventilation. Not all these plastic 'windows' open – some simply allow light into the house. They look attractive, but in summer your chickens are likely to be awake long before you are!

Design features

Doors

There should be at least one small door (pop-hole) for the chickens. There must also be a larger door so you can easily clean them out.

Some pop-hole doors pull down to form a ramp, while others slide vertically or horizontally (horizontal doors can become clogged up and difficult to use).

You can buy a device which opens and closes the pop-hole automatically (see Chapter 5). Not all doors are suitable for automation, so you may wish to consider this when choosing your henhouse.

High-rise living

Having the house raised above ground level allows air to circulate and discourages vermin from making themselves comfortable underneath your chickens. If the house is raised high enough, it provides an outdoor shelter, with space for feeding or dust-bathing.

Nest-boxes

Look for separate nest-boxes with outside access, so eggs can be collected without opening the henhouse.

If space is limited, think high rise

Nest-boxes should be in the lowest, darkest part of the house. If they are higher than the perches, the hens will roost in them, leading to mucky eggs. Sometimes nest-boxes can be closed off at night.

Perches

Chickens don't perch comfortably on broom handles! They need 5cm-square perches, with the top edges rounded off.

Allow about 25 to 30cm per bird. Although they often squash up, the lower-ranking hens may not be tolerated too close to their betters and can end up on the floor.

Low perches are needed for heavy breeds who can injure themselves by jumping down from high roosts.

Manoeuvrability

If the housing is to be moved regularly, look for a model with wheels. If possible, try it out for manoeuvrability first.

Although the house should be sturdy, remember that if it is too heavy it will be difficult to move.

Ease of cleaning

The house will need cleaning out at least every week, so the easier it is to do the job, the better.

Check the access. Imagine trying to get into every nook and cranny – and how many nooks are there to clean anyway? Perches should be removable, and there may be a droppings board which can be taken out and scraped. Sometimes nest-boxes are detachable.

Ideally the whole house should come apart for when you need to do a thorough clean.

Taking the house apart for cleaning

Henhouse Dimensions

In most cases these are the minimum recommended dimensions for standard chickens – make adjustments for tiny bantams or giant varieties. Cockerels need more room than hens of the same breed.

Floor Space in the Henhouse
At least 30cm square per chicken (assuming there is a run).

Perches
5 cm square, with the top edges rounded. Allow about 25 to 30cm of perch space per bird. Perches should be spaced at least 30cm apart (not directly underneath each other) with enough room for the tallest chickens to stand upright.

Pop-Holes
About 35cm high by 30cm wide.

Nest-Boxes
30cm square for average-size hens. Allow a nest-box for every three hens.

Run
2m square per bird if possible – a chicken should be able to stretch and flap both its wings at the same time!

Shopping around

Providing suitable living quarters will be the most expensive part of starting to keep chickens, so it's worth taking time to shop around carefully before buying. To some extent your requirements will vary depending on the types of chickens you decide to keep – in the next chapter we will look at the various breeds.

Scenarios of Keeping Chickens

In a small garden

'Keeping everything clean is the most work, but it's worth it,' says Susannah about her family's five hens. 'The chickens are pets. They recognize the sound of the car and welcome us when we come home – they also produce nice eggs.' Susannah, her husband David and sons Matthew (fifteen) and Thomas (eleven) have been keeping chickens for two years. Living on a modern estate, they were concerned whether their garden would be large enough and took time investigating the possibilities. They gained experience by helping look after friends' chickens, and found local suppliers along with trade stands at country shows to be valuable sources of information.

Their attractive patio garden is just 10 x 8.5m, but the chickens have a spacious run, covered in hardwood chips. A raised chicken house makes the best use of the space.

Trees and shrubs create shady spots, ideal for dust-baths. The result is an unexpected and delightful feature, which fits perfectly into the small garden.

The run is dug out and the ground sanitized every three months – more often in very wet weather. The 3 x 3.5m run takes three 75-litre sacks of hardwood chips to cover it adequately.

Susannah says there is no smell if the run is kept clean. Droppings are removed and sanitizer put down most days. Extra woodchips are added if the ground starts becoming muddy.

The house is cleaned with an antibacterial spray twice a week, but the droppings tray and nest-boxes are cleaned daily in summer and most days during the winter months.

The chickens are regularly wormed and are given apple cider vinegar to promote good health. They also enjoy surplus vegetables from the family's allotment, where all their waste material is composted.

In return for the care lavished on them, the hens provide around twenty eggs a week in season. These are mainly laid by the Black Rock and Light Sussex Hybrid. The three bantams (Pekins and a New Hampshire Red) only manage an egg a day between them, and stop laying altogether in winter.

'There are definite benefits to having chickens,' Susannah says. 'They all have their own personalities, and make lovely pets if handled regularly.'

In a large garden

Our garden is mainly lawn, surrounded by trees and bushes. It backs onto fields and there are no near neighbours.

The henhouse stands in a wooded area at the rear of the garden, where the ground is covered with a thick layer of woodchips and leaves. The chickens love rummaging through this under the shelter of the trees. On sunny days they sunbathe and dust-bath on a south-facing bank. They also have plenty of grass to explore and wander at will.

We have occasionally lost chickens to predators, but a colony of rooks give an advance warning and the chickens are good at hiding themselves.

At night the chickens are shut in their henhouse. This is a substantial structure, made locally from good-quality timber. We made a few alterations, including raising it 50cm above the ground, and the chickens hop up a collection of logs to the pop-holes. They took a while to get the hang of this (I found some sleeping on the back-door mat one evening), but new chickens now quickly copy the others.

Although there is little for the chickens to damage in the garden, we have recently reclaimed some space while trying to build a new terrace. Every time my husband swung the pickaxe, he would find a chicken underneath, waiting for the worms to appear. Green plastic netting has proved ideal for keeping chickens out – cheaper than wire and easier to move.

Having chickens around is fun and their eggs are fantastic – deep gold and richly flavoured from their varied diet. We keep around twenty pure-breed and cross-breed chickens, so eggs are plentiful in spring and summer but tend to disappear over winter.

Pure-breed chickens are an exotic addition to the garden. Letting them free-range keeps work to a minimum, although we have to be around to shut them in and let them out, so our lives tend to revolve around their schedules.

On a farm

Anne has kept chickens for many years, but when she and her husband Charles moved to their farm, she was able to expand.

Their previous henhouse (from the Domestic Fowl Trust) had given years of loyal service and was passed on to a daughter. Anne chose a larger model from the same company, capable of taking up to fifty birds. The hens were to free-range around a wooded area on the farm.

The first hybrid hens were purchased from the local poultry auction and egg production was soon in full swing. Unfortunately, the local fox population became interested and hens began to disappear. Eventually there was no option but to invest in electric poultry netting.

The hens now live safely behind 100m of fence, where they forage for grubs amongst the bushes and trees. When they need fresh ground, it takes Anne and Charles about three hours to move the netting.

With security taken care of, Charles presented Anne with twenty Warren hybrids one Christmas, and she now has around thirty hens in all. These include a few pure-breeds – Wyandottes, Welsummers and Marans. 'The dark brown eggs are popular,' Anne says, 'but it's a nuisance when the pure-breeds go broody. They stop laying, take up the nest-boxes and require extra attention.'

Eggs are sold from the farm gate, and Anne has no shortage of customers. She makes a small profit after paying for feed, and has plenty of eggs for her own use too. Although she has considered going into egg-production more seriously, she feels the extra work involved is too much for a busy farmer's wife.

Key Points

- Give chickens as much space as possible – don't overstock
- A small fixed run requires ground cover that can be cleaned and replaced
- Two smaller runs used in rotation are better than one large run
- Good-quality materials and careful maintenance are essential for a secure chicken run
  - If the run isn't roofed, the fence should be at least 2m high and angled outwards 50cm
  - The bottom of the fence should be dug in 50cm and turned outwards 50cm – remember the area under the door
- Movable housing has many advantages
- Unlimited free-range may not be practical in an average garden
- Free-range hens might need protection from daytime predators
- Electric poultry netting allows chickens to be moved around and keeps predators out
  - Mains power is more reliable, otherwise 12-volt batteries will be required
  - Make sure the energizer is powerful enough to run an efficient fence
  - The correct earth and proper connectors are essential
  - Keep electrified wires away from any material not insulated from the ground
  - It's important to keep vegetation trimmed and to test the fence regularly

- Correctly installed electric fencing is safe – but netting may be hazardous to some animals
- Consider the possibility of avian influenza restrictions when planning your chicken housing
- The house should be large enough for all the chickens to roost comfortably, but not so big they can't keep themselves warm
- Position the chicken house in a well-drained spot with shade and shelter
- Check poultry arks for ease of access and headroom for the birds
- A house raised above ground makes good use of space and avoids creating a home for vermin
- Cheap housing can be poor economy
- Look for solid construction and pressure-treated timber in a wooden henhouse
- In plastic housing check especially for ventilation, insulation and stability
- The house must be watertight, with a roof angled to shed water easily
- Roofing felt is a haven for red mite
- Second-hand housing may harbour disease and parasites
- Both house and run must be secure and strong enough to keep out predators
- Check for weak spots, especially around doors and nest-boxes
- Look for sturdy fixings and bolts
- Chicken wire isn't strong enough to keep out predators
- Good ventilation is essential
- There should be at least one pop-hole as well as a large door
- The house should have nest-boxes with outside access for egg collection
- Nest-boxes should be lower than the perches
- Check manoeuvrability if the house is to be moved regularly
- Make sure the house is easy to clean thoroughly

KIDS' CORNER

Quiz

There are lots of different ways of keeping chickens – how will you keep yours?

QUESTION ONE
If your garden is small, should you:
    (a) Keep as many chickens as you can squeeze in?
    (b) Have just a few chickens and give them as much room as possible?
    (c) You can't keep chickens unless you have a very large garden

QUESTION TWO
What happens to the grass in a chicken run?
    (a) The chickens scratch it up and eat it
    (b) It grows very long and has to be cut regularly
    (c) The chickens keep it short by nibbling the tips

QUESTION THREE
What would be the best place for a henhouse?
    (a) A wet area where there are plenty of puddles
    (b) In the middle of a sunny lawn
    (c) In the shade of a hedge or fence

QUESTION FOUR
Why is it important for a chicken house to be strong and secure?
    (a) To stop the chickens from escaping
    (b) To keep the chickens safe from foxes
    (c) To keep out the cold night air

QUESTION FIVE
What is a pop-hole?
    (a) The little door the chickens use
    (b) An opening for ventilation
    (c) The door to the nest-box

ANSWERS
One (b); Two (a); Three (c); Four (b); Five (a)

Did you know all the answers – or do you need another peek at Chapter 3?

Chicken Chat

'Cooped up': A chicken house is often called a coop. We talk about someone being cooped up if they are shut in a confined space. For example: 'I wish we didn't have to be cooped up in this stuffy classroom on such a lovely day!'

Chicken Jokes

What do chickens have to eat at parties?
Coop cakes!

What do you call the door to the chicken house?
The hen-trance!

Something to do . . .

Think about how you could keep chickens in your own garden.

Find out as much as you can about the various types of chicken housing. If you look on the internet you will find lots of different chicken houses. Study them carefully and see which you like best. Or ask your parents to contact some of the suppliers listed at the back of this book and ask them to send you their catalogues.

CHAPTER FOUR

Birds of a Feather:
Choosing the Right Chickens for the Job

How you keep your chickens and which chickens you choose to keep is something of a chicken-and-egg conundrum. While you may need to modify your plans to accommodate a particular breed, some situations will dictate the breed of chickens that can be kept. This chapter considers some of the more popular types of chickens, and their different characteristics. The enormous variety will surprise anyone who thinks that chickens are either brown or white. Nothing could be further from the truth.

Deciding Between Pure-Breeds and Hybrids

There are over a hundred varieties of pure-breed chickens in Britain – some of them have been around for centuries. Pure-breeds have defined standards of colour, shape and character, with the ability to reproduce offspring that closely resemble the parents.

There are also several varieties of hybrid. The original hybrid layers were developed for the battery systems, and selective cross-breeding created the 'super hen', designed to produce numerous eggs in return for the least amount of food. Different types of hybrid have since been created to suit domestic environments and small-scale enterprises. The chickens we buy for the Sunday roast are also hybrids – bred to grow much faster and fatter than their forebears.

Your initial decision will probably be whether to keep pure-breed or hybrid hens. Pure-breeds offer the most variety and are often very beautiful, but hybrids are usually more productive.

What Do You Want from Your Chickens?

When choosing a breed, consider your prime reason for keeping chickens.

Eggs

All chickens lay eggs, but there is wide variation in productivity. Hybrids are the champions, although there are also good layers amongst the pure-breeds.

Examples of hybrid layers: Warren, Black Rock, Amber Star.

Examples of pure-breed layers: Leghorn, Rhode Island Red, Sussex.

Pets

If you are more interested in having some pets, look for a friendly breed that is easy to tame. Many of the hybrids make productive pets, while small chickens (bantams) are popular with children.

Examples of pet chickens: Warren, Orpington, Silkie, Pekin Bantam.

Ornamental Birds

Look amongst the pure-breeds for colour and unusual features.

Examples of ornamental chickens: Silkie, Brahma, Cochin, Belgian Bantams, Sebright.

Eggs and Meat

Choose a dual-purpose breed for laying abilities and the requisite plump shape to provide a good meaty carcass.

Examples of dual-purpose breeds: Rhode Island Red, Sussex, Plymouth Rock.

All-purpose

What about a chicken to meet all requirements: eggs, calm disposition, attractive to look at and capable of hatching chicks that will grow into good-sized table birds?

Examples of all-purpose breeds: Wyandotte, Sussex, Faverolles.

Free-Range

Some chickens are suited to free-ranging and will find much of their own food.

Examples of free-range breeds: Appenzeller Spitzhauben, Dorking, Welsummer.

Confined to a Run

There are chickens that will adapt to living in a run, while feather-footed chickens are best kept in clean and dry conditions.

Examples of chickens for a run: Warren, Rhode Island Red, Sussex, Silkie, Cochin, Booted Bantam, Pekin Bantam – or if space is really tight consider the Serama, the smallest chicken in the world!

Hybrid Laying Hens

The hybrid has several advantages over the traditional pure-breed and is often a good choice for beginners.

As well as being productive, hybrid hens are bred to be docile and are usually friendly birds, easy to tame. Unlike many pure-breeds they should keep laying over the winter and rarely go broody.

Generally cheaper than pure-breeds, hybrids are available throughout the year and will have been vaccinated against several diseases.

Hybrids come in a selection of colours and there are varieties that lay brown, speckled or even blue eggs. They tend to be medium-sized birds that fit comfortably into standard accommodation.

Considerations when choosing hybrids

- Increased productivity comes at a price. Hybrids come into lay at around twenty weeks, and continue laying well for their first year or so. Then output starts decreasing and may cease altogether after three or four years.
- While some hybrids may still be producing an occasional egg at a grand old age, hybrids don't usually live as long as pure-breeds.
- The early and intensive start to egg production can leave hybrids vulnerable to laying difficulties.
- Hybrids can't match the pure-breeds for glamour. They don't have interesting features such as muffling, beards or feathered legs – but then these features can require extra care.
- Although hybrids are generally docile with people, they may bully their companions in a mixed flock, especially if it includes some timid breeds.
- Cockerels aren't usually obtainable and hybrids don't 'breed true' – if you breed from hybrid hens, the result will be cross-bred chicks.

Selecting Your Hybrids

Many hybrids are based on top traditional egg-layers, the Rhode Island Red or White Leghorn, with other breeds being added to produce interesting colours of both hens and eggs. The same types of hybrid may be sold under different names, depending on the breeder. These are some of the most common but new varieties are always being developed, and you may come across several others.

The best-known hybrids are the light-brown hens used by commercial egg producers. The **ISA Brown** can be expected to lay over 300 eggs in its first year (ISA stands for Institut de Sélection Animale – the French company that developed them).

Other hens of this type are the **Warren, Goldline, Gold Star, Lohmann** and **Gingernut Ranger**. They aren't exclusive to the egg industry and are widely available.

These hens are ideal for the beginner, being docile, friendly and, of course, very productive! You can buy them as pullets (young hens) and there are also various organizations that re-home ex-battery hens.

The **Black Rock** is a black hybrid with chestnut feathers around the neck. Bred for free-range (and not so suited to a confined run), it is hardy and productive, often laying and living longer than other hybrids.

Black Rocks have been developed by one breeder over many years and there is only one hatchery – The Black Rock Hatchery in Scotland (see Further Reference). You may find hens from various sources being sold as Black Rocks, but unless the supplier is registered with the official hatchery, they will not be genuine. Cockerels are never released for sale.

Similar in appearance to the Black Rock are the **Rhode Rock, Bovans Nera** and **Black Star**. They are also hardy, productive layers of light-brown eggs.

The **Columbian Blacktail** is chestnut-coloured, with a black tail and sometimes neck feathers. Waitrose sells light-brown eggs from its own free-range flocks of Columbian Blacktails. This is a hardy breed, placid and a good layer.

If you fancy blue eggs, choose a **Skyline** hen (the **Columbine** is similar). These hens often sport a head tuft and colours vary through grey/blue to cream/brown. Not every hen will lay all blue eggs, but a large proportion of them do – other eggs may be pastel shades. Pretty shells come at the expense of quantity: these hens are not as productive as some of the other hybrids.

The **Speckledy** may also lay less than some other hybrids, but the eggs are often speckled brown, and the hen is attractively mottled in dark grey and white.

Laying mostly brown eggs, the **Copper Black** is a black hen with coppery neck feathers. It is based on the Marans breed (just to confuse matters, there is also a pure-breed Copper Black Marans).

The **Sussex Star** resembles the Light Sussex pure-breed, being white with black feathers around her neck. She lays light-brown or white eggs.

The **White Star** is based on the Leghorn, and is a slim white hen who lays pure white eggs. She is one of the most productive layers, but can also be rather 'flighty' (jumpy).

The **Amber Star** matches the White Star for laying, but has a much more placid and friendly disposition. She is cream with light-brown markings and lays light-brown eggs.

The **Bluebelle** is an attractive grey/blue chicken, one of the larger hybrids. She lays plum-tinted brown eggs.

Re-homing commercial laying hens

There are several organizations that rescue commercial layers. This doesn't mean they break into chicken farms in the dead of night and liberate the inmates: they try to find new homes for hens that are of no more use to the egg industry.

Some agencies make a small charge for the hens they re-home, while others ask for a donation.

Rescued hens are often in poor condition and have no knowledge of the outside world. They need extra care, but the rescue organizations are supportive and helpful. These hens have been used to a carefully controlled diet and should initially be fed similar food. Rescued hens may also need encouragement to drink from a container. Their day has been artificially controlled, so they probably won't understand when it is bedtime or what to do when the pop-hole is opened in the morning. They will struggle to manage a steep ramp.

Most hens quickly adapt and enjoy their new way of life. Their feathers grow back and fitness improves. Adopters of rescued hens usually find it very satisfying, especially as they are such tame and friendly birds.

Their trusting nature can make them vulnerable to danger, and they may be at risk when free-ranging, especially when first released.

The hens may not continue laying in industrial quantities, but once they have settled down will often lay a respectable amount. There will also be some who stop laying altogether, and it's unlikely that rescued hens will continue to lay into old age. These hens haven't been bred to become pensioners, and although a few might reach a good age, most won't live for more than another three years at best.

Although the rescue organizations don't re-home obviously unhealthy hens, sometimes hens die soon after adoption, although they appear to be well. This may be distressing for the new owners, who can at least console themselves that they have given these hens the chance of a new life.

Pure-Breed Chickens

It can be difficult enough choosing which hybrid to keep, but that's nothing to the agonies of decision when you come to choosing pure-breed chickens.

The pure-breeds offer a vast choice, from the practical to the downright flamboyant. You can have tiny bantams or gentle giants. Some are cute and cuddly, while others are cold and aloof. You will find chickens with amazing hair-dos, beards, whiskers or feathered legs and feet. Most breeds come in a variety of colours too.

Cockerels are easily available, allowing you to breed and improve your stock. Some pure-breeds are rare or endangered, so you may like the idea of supporting and preserving these chickens. They also make an interesting talking point, especially if kept in the garden – a change from discussing the roses, perhaps?

Here are some considerations when choosing pure-breeds:

- While they can't usually match the hybrids for laying prowess, many of the pure- breeds can certainly perform well. If eggs are important, you should choose your breed carefully and try to find a productive utility strain – one that hasn't lost some of its laying abilities due to selective breeding for showing.
- Only a few pure-breeds lay in winter, while some will go broody in spring, which stops egg production.
- Pure-breeds tend to continue laying for more years than hybrids, and may live longer too. My Leghorn hen is at least seven years old and still produces an egg every other day in season – but she takes a long rest over winter too.
- Pure-breeds can offer other services apart from eggs. Broodiness can be an advantage if you want to hatch some chicks, and many of them are good meat birds.

Fancy features

There are many attractive pure-breeds that require no more maintenance than any other chicken, but special features will require extra attention:

- Birds with heavily feathered feet should be kept in dry conditions to avoid a build-up of mud between their toes. These chickens are reputed to be less likely to dig up the garden – although I had a Pekin Bantam who could make a mean dust-bath, and my big Cochin cockerel loved to make himself comfortable in the flowerbed.
- If 'low-slung' chickens with fluffy feathers close to the ground are to free-range, they will need a clean, well-ventilated henhouse to dry off overnight.
- Crests and beards should be regularly inspected for mites. A narrow-lipped drinker will be necessary to keep feathers from becoming wet and soiled.
- Large combs and wattles can be at risk of frostbite – rubbing in Vaseline helps to protect them.

Picking Pure-Breeds

Some pure-breeds are not really suitable for beginners, or can be rare and difficult to find. Entire books have been compiled on chicken breeds alone, but this is a selection of those breeds that are commonly available and likely to be of interest when starting out.

The pure-breeds are divided into different classes; understanding these will help when deciding which birds are best suited to your needs.

Hard feather, soft feather

Chickens may be either hard- or soft-feathered.

Hard feather refers to the closely fitting plumage of the 'game breeds', which were originally bred for fighting.

Now that cock-fighting is illegal, game breeds are kept for showing and still have a devoted following. A particular quality of these birds is 'attitude', and they will tend to reflect their fighting origins: the hens can be aggressive with each other and the males may fight to the death. Mixing them with other breeds isn't advisable, and they are best kept in small groups. The game breeds are not the greatest layers but make very protective mothers. Although pugnacious amongst themselves, game birds are usually easy to tame and can become extremely attached to their owners – but they aren't particularly suited to the beginner.

Game breeds include: Old English Game, Modern Game (especially bred for showing), the Asil and the Shamo (both Asian breeds).

Soft feather refers to all the other chicken breeds. Their plumage is looser and sometimes very fluffy.

Soft-feathered chickens are separated into light, heavy and bantam breeds.

Light breeds

Light breeds are layers rather than table birds. Many light breeds, especially those of Mediterranean origin, can be flighty. They may be more difficult to tame and can also be adept fliers, although they are less likely to go broody.

This doesn't apply to all light breeds – the Silkie, for example, is a docile non-flyer and a committed broody.

Popular light breeds include:

Araucana

This breed with its bushy facial feathering is a good layer of blue eggs. Sometimes called the 'Easter Egg Chicken', it lays eggs whose colour may range from green to khaki or pink (although variations indicate the breed strain isn't pure). Araucanas are hardy and fairly placid – they can free-range or live in a run. The most popular colour is lavender (pale grey).

Cream Legbar
This is another layer of blue eggs, thanks to some Araucana blood in its make-up. The Cream Legbar is an auto-sexing breed – male and female chicks are different colours when hatched. The colour of the adult birds is mainly cream and grey, and there is a small crest. They are active chickens that may require some extra effort to tame.

Leghorn
This is one of the prime layers, and has been used in creating many hybrid varieties. A slim bird, it lays large white eggs and comes in a wide variety of colours. Although it can be kept free-range or in a run this breed is often flighty, and while it can be tamed the Leghorn doesn't usually appreciate being handled.

Silkie
After his Asian travels in the thirteenth century, Marco Polo wrote about chickens 'which have hair like a cat, are black, and lay the best of eggs'. Silkies now come in several colours, and are popular as pets due to their friendly nature and feathers that resemble fur. These strange feathers aren't very efficient at protecting them from cold and wet, so they would be a good choice for a covered run. Their furry topknots can also harbour mites. Silkies aren't particularly good layers, but are enthusiastic broodies. They are susceptible to Marek's disease, which causes paralysis and death. Check that the birds have been vaccinated before buying.

Welsummer
This breed is known for its large, dark-brown (often speckled) eggs. One of the larger light breeds, the Welsummer is an attractive, traditional-looking chicken. Welsummers are hardy birds, capable of supplementing their diet by foraging if free-range, but they can also be kept in a run. This is one of the calmer light breeds, although some strains are livelier than others.

Fancy something a little different?
You might also consider the *Appenzeller Spitzhauben*. This old Swiss breed can be skittish, so perhaps it is not entirely suitable for the beginner, but if allowed to free-range will happily find much of its own food. It will take patience to tame this chicken, but it is particularly pretty with its forward-pointing crest of

feathers – Spitzhauben means 'pointed bonnet' in German. The most popular colour is 'silver spangled' (black spots on a white background). Small and sprightly, the Appenzeller produces a medium-sized white egg.

There are many other light breeds – some of them very flighty. Although I find their independent natures rather attractive, these lightweight chickens don't appeal to everybody. If you want a cuddly companion or prefer a more matronly type, then consider a heavy breed.

Heavy breeds

Generally more placid than the light breeds, these hens are also more likely to go broody. Several of the heavy breeds are prolific layers, and were traditionally kept for both eggs and meat.

Popular heavy breeds include:

Brahma

The Brahma is a really heavy – and large – breed. These chickens will require extra space in the house, on the perches and through the pop-holes. They are stately, docile birds, easy to tame and well suited to be garden pets, although their size can make them difficult for children to manage. Brahmas have feathered legs and feet, and shouldn't be kept in wet conditions. They don't fly, and usually get on well with other breeds, although they can sometimes be bullied. This breed matures slowly, so you may have to wait a while for their first eggs, which are not very large considering the size of the hens. They aren't particularly prolific layers and the hens are likely to go broody.

Cochin

Even heavier than the Brahma, the Cochin is also a placid breed and makes an excellent pet. Breeding for showing has reduced its former egg-laying abilities, but it is a handsome bird with profuse feathering. The big feathery 'trousers' and feathered feet will require clean and dry conditions, but these birds can live happily in a run as they are not particularly active. However, a too-sedentary lifestyle may lead to obesity and heart problems. The hens make good broodies.

Dorking

The Dorking is reputed to have come to Britain with the Romans. Although originally bred for the table, it is also a reasonable layer. This breed prefers to free-range but is docile, and if handled properly can become tame. With a long body and short legs, the Dorking is unusual in having five toes – most breeds have four.

Faverolles

This French breed has Dorking amongst its ancestors and also has a fifth toe. A much-bearded chicken (I call Faverolles 'Father Christmas chickens'), it is usually an attractive salmon colour, although other colours may be found. Prized for its meat as well as being a good layer, the Faverolles would make an attractive addition to the garden. They are friendly, affectionate chickens which are easy to tame, but may be bullied by other breeds.

Marans

Maran eggs are prized for their thick, glossy, dark-brown shells. The shells have smaller pores than usual, which is thought to help avoid contamination. Marans are good layers and given the chance will happily free-range. They can also be kept in a run, but easily become fat. These are friendly birds and they can become reasonably tame.

Orpington

The buff Orpington was a favourite of the late Queen Mother, and buff is perhaps the best-known colour, although there are many others. These attractive chickens have short legs, which can hardly be seen under their masses of feathers. They enjoy free-ranging and are well suited to living in the garden. Their friendly and affectionate nature makes them excellent pets – owners often become very attached to their Orpingtons. Breeding for showing has reduced their former laying ability, but they tend to go broody and can cover large clutches of eggs. Their gentle nature can lead to them being bullied in a mixed flock.

Plymouth Rock

A popular dual-purpose breed, especially in the United States, these are placid chickens and easy to tame. They appreciate being able to free-range. The 'barred' colour scheme is best known – a striking pattern of black and white stripes. This breed has excellent laying abilities and features in the creation of several hybrids.

Rhode Island Red

This famous dual-purpose breed is one of the top layers of light-brown eggs and has been widely used in the development of hybrids. They mature early, can be kept free-range or in a run and are generally placid, friendly birds. The colour is a deep reddish-brown.

Sussex

This dual-purpose breed comes in a good variety of colours, but the Light Sussex (white with a black neck and tail) is the best known. The Light Sussex is considered the most productive, but all colours should lay well. These birds can either free-range or will adapt to life in a pen – they are also easy to tame. This all-round chicken would be an excellent choice for the beginner.

Wyandotte

Another good breed for the beginner, the Wyandotte comes in a large choice of colours. It makes an attractive addition to the garden – the silver-laced (white feathers outlined with black) is particularly eye-catching. This breed can live free-range or in a run, has a placid nature and is usually easy to tame. These are fairly large dual-purpose chickens with a pleasant, rounded shape, made to look larger by abundant feathering. They are excellent layers, although very likely to go broody. The white Wyandotte is the utility strain and the most productive.

Many heavy breeds are ideal all-round chickens, but if space is limited or you need a pet for small children, then consider the bantams.

Bantams

These are small chickens. They may be either scaled-down versions of large breeds or 'true bantams', meaning they have no large equivalent.

Most of the light and heavy breeds mentioned earlier have a readily available bantam counterpart. Some lay as well as or better than their big cousins, but their eggs are smaller.

The true bantams aren't the best of layers but can make lovely pets, being small, pretty and usually very friendly (although some males can be feisty). Those with feathered legs and feet are better suited to an enclosed run, although they will appreciate a chance to roam around the garden. Some can fly well. The smaller birds will be at risk from a variety of predators.

There are several breeds of true bantams – these are some of the most common:

Belgian Bearded Bantams

There are three main varieties of these charming little birds.

The Barbu d'Anvers (or Antwerp Bearded Bantam) has generous muffling and beard but no leg feathers. It comes in many colours, lays a fair number of white eggs and is likely to go broody.

The Barbu d'Uccle is also bearded and muffled, but has feathered feet and legs. It too comes in a range of colours, the most popular being millefleur. This consists of deep orange feathers with a black and white pattern – as the name suggests, like a thousand flowers. The d'Uccle also lays a reasonable amount of white eggs and is likely to go broody.

The Barbu du Watermael is the smallest, being similar to the d'Anvers but more delicate-looking. It comes in a vast range of colours and makes a good pet, but doesn't lay many eggs and frequently goes broody.

Booted Bantam

Another pretty little chicken with abundant leg and feet feathers, this bantam is also known as the Sabelpoot and is available in a large choice of colours. It makes an ideal pet for children, but isn't a great layer and often goes broody.

Pekin Bantam

These cuddly bantams are one of the most popular breeds for children. They seem to enjoy human company but can be surprisingly feisty with other chickens, fearlessly taking on much larger breeds. Pekins are widely available in many colours. Their fluffy feathers come down to their feathered feet so they are better kept in dry conditions, although they appreciate a chance to forage in the garden. They lay a reasonable number of eggs and are enthusiastic broodies.

Rosecomb Bantam

Mainly kept for showing or as pets, the defining feature of this attractive bantam is the 'rose comb' – a flat knobbly comb, finishing in a spike. They have distinctive white ear lobes and carry their wings elegantly low. These bantams are easy to tame and enjoy roaming around (they are good fliers), although they can also be kept in a pen. They are less likely to go broody than some bantams, but their eggs are small and they don't lay very many.

Sebright

This is a particularly eye-catching bantam, available as silver- or gold-laced (white or gold feathers with black borders). They are delightful characters but won't supply many eggs. They enjoy foraging in the garden but are able to fly well. This breed is susceptible to Marek's disease, so make sure they have been vaccinated before buying. They are difficult to breed successfully and only go broody occasionally.

Serama

This breed is becoming more popular in the UK, and is included because it is the smallest breed of chicken in the world, so may be suitable if space is limited. In their native Malaysia, Seramas are sometimes kept in the house; although fairly hardy, they need some protection in cold weather. Eggs are tiny (five are equivalent to one standard egg) and not very plentiful. Hens will occasionally go broody. They are confident, friendly birds, easy to tame. The breed club website says the Serama is often described as a 'living work of art'.

Chickens Don't Always Follow the Rules!

There are so many different types of chickens to choose from that it should be easy to find one that will exactly suit your requirements – except chickens can break the rules whenever they like.

Although chickens usually conform to their stereotypes, there can be variations, so you should be prepared for the unexpected. 'Shadow', my friendly Leghorn, was always under my feet, although her sister displayed the normal aloof nature of this breed. Obnoxious Orpingtons, non-sitting Silkies and easy-going game fowl probably exist too.

When my Appenzeller (one of the lightest, flightiest, non-sitting breeds) disappeared, we feared the worst. She was discovered sitting on a nest of twelve eggs, all of which she hatched and raised as proudly as any Mother Hen. True, she didn't tolerate any slackers, and the chicks grew up to be tough, independent little birds, keen foragers, who insisted on roosting in the trees. All of which are Appenzeller characteristics . . .

Finding Out More

Once you have decided which types of chickens are likely to be suitable, try to find out some more about them.

Pure-breeds

There is a dedicated breed club for most breeds (those without a club are looked after by the Rare Poultry Society). Contact the relevant club for information, advice and help with finding breeders. The Poultry Club of Great Britain has contact details for breed clubs (see Further Reference).

Hybrids

Visit one or two suppliers to look at the different varieties they have available. The breeder will be able to advise you on their characteristics and suitability for your requirements.

Seeing before you decide

A poultry encyclopaedia or internet search will provide useful pictures and information but it's better if you can see the actual birds.

Country shows often include poultry exhibits, or you could attend one of the many poultry shows, which are held at both local and national level. A large event will have a vast selection of chickens, as well as enthusiasts keen to talk about their favourite breed. The Poultry Club has details of shows throughout the country. Well worth a visit are the National Poultry Show and the Federation of Poultry Clubs Championship Show ('the Fed'). Held in November/December, these major events are the highlight of the poultry exhibitor's year and attract hundreds of entries.

While shows tend to focus on pure-breeds, a visit to a large poultry sale will provide an opportunity to view all kinds of chickens. Don't expect to see the best examples, as breeders often use markets to sell surplus stock – and it's not a good idea to buy your first chickens from sales either (see Chapter 7).

Key Points

- Chickens may be either pure-breeds or hybrids
- Hybrids are productive, usually friendly and easily available
- Productivity in hybrids declines after the first year or so
- Rescuing ex-battery hens can be very satisfying, but they require extra care at first
- Pure-breeds offer a wide choice of attractive chickens, and cockerels are available
- Many pure-breeds stop laying in winter, but often lay for more years than hybrids
- Features such as abundant feathering require extra attention
- Hard-feathered breeds are the game birds, originally bred for fighting
- Soft-feathered applies to all the other breeds
- Light breeds are bred for laying rather than the table
- Several heavy breeds are prolific layers and can also be kept for meat
- Heavy breeds are generally placid – some tend to go broody
- Bantams are either small versions of large chickens or 'true bantams' with no large equivalent
- Find out about your chosen breed, but remember that chickens don't always follow the rules!

KIDS' CORNER

Quiz

Do you know which chickens you want to buy yet? Try this quick quiz on Chapter 4.

QUESTION ONE

Which of these statements is true?
- (a) All chickens look the same
- (b) Chickens only come in brown or white
- (c) There is an enormous variety of chickens to choose from

QUESTION TWO

If you want eggs in winter, which type of chickens would you choose?
- (a) Hybrid hens
- (b) Pure-breed chickens
- (c) Chickens don't lay eggs in winter

QUESTION THREE

Do ex-battery hens make good pets?
- (a) No, they are very unfriendly birds
- (b) Yes, they quickly become tame
- (c) Battery hens aren't available to the public

QUESTION FOUR

What are 'wattles'?
- (a) The red object on top of the chicken's head
- (b) The flesh dangling under the beak
- (c) Extra feathers around the face

QUESTION FIVE

Is a bantam:
- (a) A small chicken?
- (b) An extra-large chicken?
- (c) A chicken with feathered feet?

ANSWERS

One (c); Two (a); Three (b); Four (b); Five (a)

You should know a lot about the different types of chickens now – did you get all the answers right?

Chicken Chat

'As rare as hens' teeth': Chickens don't have teeth! This saying is used about something that is very unlikely to exist or impossible to find. For example: 'I've been searching the shops all day, but trousers that will fit me are as rare as hens' teeth.'

Chicken Jokes

Do you know which side of a chicken has the most feathers?
The outside, of course!

Why are chickens so well groomed?
Because they always have a comb!

Something to Do . . .

Think about which types of chickens you would like to keep. Make a note of some suitable breeds or hybrids. Now try to find out more about them. Look on the internet or ask your parents to contact the breed clubs. Ask your friends about the chickens they keep and try to visit a poultry show or sale.

CHAPTER FIVE

Buying the Eggstras:
What Else Will You Need?

You've selected an ideal home for your new chickens, and have decided on the ideal chickens to come and live in it. Surely there's nothing else to do now but reach for your wallet?

You might just about manage without anything extra, but having the right equipment will make life easier for you and better for the chickens.

There is a vast selection of chicken-related items available online, or you can send for catalogues. Alternatively, agricultural stores usually carry a reasonable selection of poultry equipment.

These are the basics you will need.

Choosing a Poultry Feeder

A proper feeder avoids waste. Chickens are messy eaters and will happily leap into an open dish, scattering food and adding a couple of droppings in the process. Spilt food soon becomes stale (which can make chickens ill). It also attracts vermin, and if it gets wet will smell very unpleasant – worse than anything the chickens can produce!

There are various styles of feeder available – these are some of the most popular.

Cylinder feeders

This popular design is readily available in a variety of sizes. The food is held in a central cylinder and trickles into the surrounding trough, which may be divided into partitions – useful for energetic eaters.

Check there is space under the cylinder for the food to flow easily into the trough – some models are adjustable.

The cylinder should have a lid and may feature a circular 'rain cover'. This helps keep the food in the trough dry and is essential for outdoor birds. Be wary of designs that have a central rod passing through the cover, as these can leak in heavy rain.

Raising the feeder above ground level helps prevent the chickens from scratching out the food and keeps it free of debris too. Some feeders have legs, and you can also buy feeder stands – or be self-sufficient and use a couple of bricks.

A hanging feeder keeps food well off the ground and also away from rodents.

The 'Blenheim' feeder features a partitioned trough, rain-hat and legs

Trough feeders

Open troughs can only be used under cover – although some designs feature a lid. Troughs with partitions help prevent chickens from scratching out the food.

Peck and treadle feeders

The peck feeder consists of a container on legs with a spring inserted underneath. The chickens release the food by pecking at the spring. You can buy these complete or purchase a spring to make your own.

The 'Brittany' feeder can be hung up, and also has a rain cover

The treadle feeder is a box-like device that opens when the chicken stands on a metal plate.

Peck and treadle feeders protect food from wild birds and vermin, but are generally larger and more expensive than cylinder feeders.

An open trough with partitions

Feed cups

These hook on to wire mesh and can be used to contain grit or supplements. They also make handy individual feeders/drinkers if birds have to be segregated.

Feed cups clip on to wire, and have many uses

Metal or plastic feeders

Metal feeders are usually galvanized, although some are enamelled steel. Metal is more expensive but more robust than plastic. Lightweight plastic feeders may be blown or knocked over and damaged. Chickens can give their feeder some abuse – my cockerels balance on top of the feeder and rock it until it falls over.

What size feeder?

Buying an extra-large feeder to save having to refill it every day can result in stale or mouldy food. However, the feeder should be big enough

A 7kg galvanized feeder

to contain enough food to last the flock at least a day. An average-sized chicken needs roughly 120 to 150g of food daily, but the senior chickens will eat more (because they can). You will need to feed above the minimum to ensure the lower orders get their share.

Providing extra feeders can help prevent squabbles and bullying.

Grit hoppers

Chickens need flint grit to enable them to process their food (see Chapter 6).

You can buy a grit hopper, or use a small feeder or a little dish.

Feed scoops

A scoop for transferring food from bag to feeder is handy though not essential – you could use a cup.

Choosing a Poultry Drinker

A poultry drinker prevents water from being spilt and helps keep it clean.

Tower drinkers

The water is contained in a tower and runs into a surrounding trough. In many models the tower part is turned upside down to be filled, then the base is attached and the whole lot turned over – with large drinkers this can be a messy business, leading to wet feet. It can be worth paying extra for a drinker that is filled from the top if catering for several chickens.

An 'Easyfill' tower drinker

Some of the cheaper plastic drinkers don't last very long – and the tower usually pushes directly on to the base, with only one hole for the water to escape. If that becomes clogged up, the chickens will go thirsty. Better designs have a base that locks into position.

If the drinker doesn't have legs or a hanging loop it should be placed on a stand or bricks to avoid the water becoming soured with debris.

Chickens with beards, muffs and crests need a narrow-lipped tower drinker to prevent their feathers from becoming soiled.

Bucket drinkers

This type of drinker resembles a bucket lying on its side. A plate over the wide end keeps the water in, with a trough for the

Bucket drinker

chickens to drink. The design keeps the water clean and out of the sunlight – it has had some good reviews.

What size drinker?

Chickens drink a surprising amount of water. An average laying hen will get through at least 500ml a day. The drinker should hold more than enough water to last the flock for a day, or you could supply extra drinkers to avoid arguments (water can be guarded almost as zealously as food). Whwn choosing a drinker, bear in mind that water will evaporate more quickly from small drinkers in hot weather.

Metal or plastic drinkers

Drinkers can be plastic, galvanized or enamelled steel. Some metal drinkers have glass tops so you can check the water level easily. In plastic drinkers the water is visible through the white plastic tower.

Metal drinkers are more robust, but can't be used with apple cider vinegar. If adding this supplement to the drinking water, you will need a plastic drinker. However, plastic drinkers can crack, especially if the water becomes frozen.

All drinkers should be regularly cleaned to prevent a build-up of poisonous algae. White plastic and glass-topped drinkers will need cleaning more frequently. Placing drinkers out of direct sunlight helps prevent algae forming, and keeps the water cool as well.

A 2.5-litre galvanized metal drinker   A 2.5-litre glass-topped metal drinker

Plastic 'tower drinker' with locking base

Some drinkers have legs . . .

. . . or the drinker can be placed on bricks to keep the water clean

Storing Feed

News of a free meal spreads quickly in the rodent community, and they will find a way into your storage area if they scent food. They have a very good sense of smell – I left some corn in the boot of my car and somehow Ratty sniffed it out. We never discovered how he got in, but for weeks afterwards I had visions of a whiskery face peering at me through the air vents.

Small feed storage bin

So store your feed carefully and sweep up any spillages. Containers need to be secure and airtight – damp feed quickly goes mouldy and will have to be thrown away. Keep your feed container under cover, for example in a shed or garage.

You can buy proper storage bins, but unless you need to keep large quantities of feed, a dustbin is cheaper and will take a 20kg feed bag with room to spare. If you have several chickens an extra bin will allow you to store a reserve bag, but chicken feed has 'use by' dates so it's not a good idea to stockpile.

Metal bins are more expensive but more secure. Rats have been known to chew through the plastic variety – although unless you have a real rodent problem, a heavy-duty plastic dustbin should do the job. Make sure the lid fits snugly, and stand the bin above ground level if there is a possibility of damp.

A spare bin is useful for storing the chicken feeder overnight. Left-over food shouldn't be returned to the feed bag, but if it hasn't got damp it can be used the next day. Always remove the feeder at night – even if empty it will attract vermin and their droppings can spread disease.

Providing a Dust-Bath

You should provide a dust-bath if your chickens aren't able to make one – or if you wish to dissuade free-ranging chickens from making one in the garden.

A shallow plastic storage box, a sandpit tray or even an old tyre could be used. A lid keeps the soil dry in wet weather and prevents cats using it at night. Cardboard quickly disintegrates and the chickens will destroy it.

The box should be two-thirds filled with fine, dry soil or play sand – don't use sharp sand as this can irritate the chickens' eyes. Wood ash mixed into the soil will make it lighter.

Some poultry equipment stores sell dust-bath trays and suitable sand. Place the dust-bath in a dry area, away from food and water containers.

Make sure other animals can't use the dust-bath as a toilet and that children don't mistake it for a sandpit.

You can make or buy a dust-bath – this one has its own cover . . .

. . . and can also be used to keep feed dry

Giving Chickens Some Fun in their Run

Chickens enjoy perching during the day, and some runs come equipped with perches. Otherwise you could add some branches or logs.

You can even buy a 'chicken gym' – complete with ladders, perches and hooks to hang treats.

Keeping confined chickens occupied helps prevent anti-social behaviour.

Selecting Bedding

Chickens don't usually sleep on the floor, but bedding material or 'floor litter' absorbs moisture, adds some insulation and makes cleaning easier. The nest-boxes should be well-lined with soft bedding to make them attractive to the hens and protect the eggs. You will need enough bedding to cover the floor of the henhouse to a depth of about 2cm, with a thicker layer for the nest-boxes.

The most important points to consider when choosing bedding are its absorbency and dust-free properties. Dampness and dust cause health problems in chickens.

Dust-extracted shavings are a popular choice but shredded hemp, which is often used in stables, is another option. It's more expensive than shavings but four times more absorbent, and some brands have eucalyptus fragrance added. There are also various bedding materials that have been developed especially for chickens. Shop around and see what suits your circumstances and budget.

Although straw is sometimes used, it is dusty in the confines of a coop and can harbour mites. It is also less absorbent than shavings, and should be turned over regularly to prevent it becoming compacted. Chopped straw is better and is sold dust-extracted for poultry. Don't use hay or sawdust. Hay quickly compacts and can harbour harmful spores and moulds. Sawdust is – very dusty! It can cause eye irritation and respiratory problems.

Shredded paper tends to blow about and can be messy. It isn't very absorbent and quickly becomes compacted. Shredded cardboard as sold for animal bedding would be a much better choice. It is more absorbent than paper, is dust-free and composts easily.

Large bales of bedding are usually more economical if you have somewhere dry to store them.

Keeping Things Clean

Equipment
A small shovel and a large bucket are the basic requirements, along with a stiff brush. A scraper is ideal for removing cement-like droppings. You can buy these items from poultry or hardware stores – or you may already have them at home. A washing-up brush kept by the tap makes it easy to clean feeders and drinkers, while heavy-duty rubber gloves are useful for dirty jobs. Soiled bedding makes great compost – keep it out of reach of the chickens in a compost bin.

Cleaning materials
Powder disinfectant helps dry up wet patches and keeps down infection.

If your chickens are to live in a fixed run, add a DEFRA-approved poultry disinfectant and some ground sanitizer to your shopping list.

For scrubbing the henhouse, Poultry Shield is an effective cleaning agent, and also helps combat red mite.

Diatomaceous earth is a natural weapon against insects and parasites. It can be sprinkled around the henhouse, nest-boxes, and dust-baths, applied directly to the birds, and even added to their feed. Make sure you buy 'food-grade' diatomaceous earth, preferably a product sold specifically for poultry, and refer to the manufacturer's instructions before using.

All these products can be bought from poultry supply stores.

Putting Together a First Aid Kit

There are a wide range of medicinal products available, but to start with a basic kit should be adequate. Many items lose their effectiveness over time so are better purchased as required. Keep your first-aid essentials in a labelled container where they are handy. You might include the following:

- Antiseptic spray
- An approved poultry disinfectant (e.g. Virkon)
- Surgical spirit
- Petroleum jelly
- Eye-dropper or syringe (for administering fluids)
- Scissors
- Tweezers
- Cotton wool
- Gauze pads
- Duct tape
- 'Vetwrap' bandage (used on horses – sticks to itself, not to hair or feathers)
- Disposable gloves
- A couple of small towels or old face flannels

This is a minimum list, which can be extended according to your own circumstances and experience. See Chapter 11 for further information on chicken ailments and problems.

Using Automated Equipment

A useful device will electronically open and close the pop-hole (some types of pop-hole door aren't suitable for automation). These systems either operate according to light levels or are used with a timer.

Chickens go to bed at dusk, so will be roosting by four o'clock in winter. If the pop-hole isn't closed, they will be vulnerable to predators. The opposite applies in summer, when they might not be in until 10 p.m. – which may complicate your social life.

The automatic device will also let the chickens out in the morning if you fancy a lie-in.

A disadvantage of an automatically closing pop-hole is that a chicken could be shut out of the house – there's always one who leaves it until the last minute to go to bed. One of our hens got stuck in a pile of railway sleepers and would have remained there if we hadn't done the usual head count. Some hens make nests outside and settle down on them.

The possible loss of one or two chickens may outweigh the chances of losing the lot, but an automatic pop-hole should be seen as a helpful assistant rather than a replacement poultry keeper. Before settling down for the evening, you should check that all your chickens have come home to roost.

There are also automatic feeding and watering systems, but these tend to be more suitable for large flocks. They can be useful, but should never be allowed to replace the daily care and checks that form an essential part of keeping chickens.

Key Points

- Poultry feeders help to avoid waste
- Feed must be kept dry
- The feeder should be big enough to contain sufficient food for at least a day
- Metal feeders and drinkers are more robust, but more expensive than plastic
- Some of the cheaper feeders and drinkers may be poorly designed or rather flimsy
- Raise feeders and drinkers above ground level
- A poultry drinker prevents water from being spilt and helps keep it clean
- The drinker should be large enough to contain plenty of water – a hen drinks about 500ml a day
- Metal drinkers aren't suitable for apple cider vinegar
- Clean drinkers regularly to prevent a build-up of poisonous algae
- Store feed under cover in secure containers
- A plastic box with dry soil or sand is ideal for a dust-bath
- Keeping chickens occupied in the run helps prevent anti-social behaviour
- Bedding should be dust-free and absorbent
- A shovel, bucket, stiff brush and scraper will be needed for cleaning the henhouse
- Poultry stores can supply a good range of products
- Keep a basic first-aid kit at hand
- Automated equipment is useful but cannot replace daily care

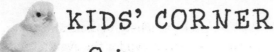

KIDS' CORNER

Quiz

Find out how much you have learnt from Chapter 5.

QUESTION ONE

What's the best way to stop chickens from wasting their food?
    (a) Put it into a dish or bowl
    (b) Use a chicken feeder
    (c) Scatter it on the ground

QUESTION TWO

How much water can a hen drink in a day?
    (a) 15ml
    (b) 50ml
    (c) 500ml

QUESTION THREE

How should chicken feed be stored?
    (a) In a strong, secure container
    (b) In a cardboard box
    (c) In a plastic bag

QUESTION FOUR

Which of these statements is wrong?
    (a) Feeders and drinkers should be placed above ground level
    (b) Poisonous algae can grow in drinkers that are not cleaned regularly
    (c) Chickens don't perch during the daytime

QUESTION FIVE

Which of these materials is best for chicken bedding?
    (a) Hay
    (b) Straw
    (c) Wood shavings

ANSWERS

One (b); Two (c); Three (a); Four (c); Five (c)

How many answers did you know? Look at Chapter 5 again if you are unsure about anything.

Chicken Chat

'To over-egg the pudding': This means to spoil something by trying too hard to make it better. For example: 'Daniel bought so much chicken equipment that there was no room left in the run for the hens – he over-egged the pudding!'

Chicken Joke

Why did the chicken join the band?
Because he had the drumsticks!

Something to Do . . .

Make a list of all the equipment you will need for your new chickens. Look in the suppliers' catalogues or on the internet to see the different styles that are available. Compare the quality and prices.

CHAPTER SIX

Chicken Feed:
Organizing the Daily Rations

These days it's very easy to feed chickens. Ready-made feeds have been formulated to contain just about everything a chicken needs for a healthy egg-laying life.

Feeding Scraps

In the 'old days' chicken feed was regularly supplemented with table scraps. Even potato peelings were boiled up for the chickens.

Since the foot-and-mouth outbreak in 2001, it has been illegal in Britain to feed kitchen waste to any chickens – including those kept as pets. This causes a fair amount of debate amongst chicken keepers, as many cannot see the point of it.

The law

The law is intended to prevent the spread of disease through food-producing animals being fed anything that could have come into contact with products of animal origin.

Nothing that has gone through the kitchen should be fed to chickens, because there is a risk of cross-contamination. You can feed vegetables straight from the garden, but not if you have first prepared them in the kitchen.

Even garden vegetables grown on kitchen-based compost shouldn't be given, and the chickens shouldn't have access to the compost heap either.

Phew – you probably didn't realize your kitchen was such a hotbed of infection! Only vegan kitchens are exempt.

While it's unlikely that someone from DEFRA will pop up just as you throw the chickens a few crusts, it's as well to know the rules and the implications of breaking them. Substantial fines can be imposed, and should there be an outbreak of a major disease, many chickens will die while many others will be culled. This applies to back-garden chickens too, as they could potentially pass on diseases.

Commercial flocks are regularly tested for salmonella – it's especially important to stick to the laws on feeding if you are intending to sell eggs.

Other reasons for not feeding scraps

Much of the food we eat contains additives and ingredients that are unsuitable for chickens and could make them ill.

Chickens tend to fill up on 'extras' at the expense of their carefully balanced feed (like children eating sweets before dinner). Unbalanced feeding can leave chickens short of vital nutrients, which will affect health and egg quality.

Hybrid hens in particular require a good diet. If not fed correctly, they will take the necessary nutrients from their own bodies to aid their intensive egg production.

Chickens are dependent on their owners for food, and it is easy to supply them with what they need. Why make life difficult?

Buying Feed

Agricultural stores, feed merchants and some pet shops sell chicken feed, or it can be ordered online. Choose a feed that is easily available, buy from a store that has a good turnover of stock and check the 'use-by' dates on the bags.

Don't buy so much feed that it can't be used before the expiry date, but try to keep enough in reserve so you don't run out. Remember the chickens when shopping for Christmas – unlike supermarkets, feed stores sometimes close for a few days over holiday periods.

Put feed into the storage bin when you get it home, but don't mix new feed with old – leave the new bag unopened until you need it. If food is left in its bags, you can usually store a new bag under a partly used one.

Choosing feed

Complete feeds are sold for all types of chickens: layers, growers, chicks, breeding stock, free-range, etc. Some feeds are organic and some may contain chemical additives (as preservatives, to prevent disease or to make yolks yellow). There's plenty of choice, so make sure you select the correct feed for your birds.

Choose a brand from a reputable manufacturer to ensure a good diet for your chickens. Feed manufacturers often have an advisory service too.

Feeds are available as pellets, mash and crumbs.

Pellets

These don't look very exciting, but are designed as a complete chicken feed. Pellets are convenient, clean and easy to handle. Each pellet is the same, ensuring all the birds receive a balanced diet (small-size pellets are available for bantams).

Mash

Mash can be fed dry or mixed with water. Moistened mash should be fed in a trough and quickly goes sour – uneaten food must be thrown away after a few hours.

Dry mash is more difficult for chickens to eat than pellets. Working for their food keeps confined birds busy and helps prevent sedentary breeds from becoming too fat.

Chickens make a mess eating dry mash, and spilt food should be cleaned up regularly. It's likely to end up in the drinker too, souring the water and entailing frequent refills. Mash isn't suitable for chickens with beards or crests as it becomes stuck in the feathers, which can lead to attacks from other birds.

Crumbs

Crumbs are usually sold for chicks, although they are now being produced for ex-battery hens, who have only ever known crumbed feeds.

Supplying Grit

How a chicken digests its food

As chickens don't have teeth, food is swallowed more or less whole and stored in the crop (a pouch at the base of the neck). The food then passes through a small stomach and on to the gizzard.

The gizzard is a muscular organ that grinds food with the help of small stones swallowed by the chicken.

Flint or insoluble grit

Free-range chickens pick up stones from the soil, but if chickens are confined or have limited free-range, they will need a container of grit in their run. Keep this topped up so they can help themselves. You'll find flint grit (often called 'insoluble grit') at the feed suppliers.

Some feeds may contain grit but it's unlikely to be enough for domestic chickens, especially if they are fed grain as an extra.

Soluble grit

Soluble grit usually consists of oyster shell and is a dietary supplement, adding extra calcium for strong eggshells. Feeds usually contain sufficient calcium, but if you provide a hopper of soluble grit, the hens will take what they need.

Mixed grit

Alternatively you can buy a mixture of the two types of grit, which should cater for all eventualities.

Providing Green Food and Grass

Chickens love to eat grass, and eating green foods makes their egg-yolks naturally golden. Short grass is better than long as it is more nutritious and long stems can cause digestive problems.

If access to grass is impossible, you should hang up some greens in the run. Hanging up vegetables prevents them from being trampled and provides entertainment for the chickens too. Try to vary what you feed – broccoli, cabbage, cauliflower, spinach, sprouts, swede and sweetcorn will all be enjoyed (lettuce sometimes causes digestive upsets and an excess of cabbage may result in diarrhoea).

You could also pick wild plants such as dandelions and chickweed. Even young nettles will be appreciated, especially if they have some insects clinging to the stalks. Make sure any plants you gather haven't been polluted by cars, pesticides or dogs.

Don't leave left-over greens to fester – clear them up each evening.

Some people grow trays of grass for their chickens – if you cover the trays with mesh the chickens can't scratch out the compost, and you can re-use it for the next sowing.

Feeding Grain

If you scatter some grain in the afternoon, the chickens will have a great time foraging for it, and the warmth it generates will keep them cosy at night. Feed about one handful per hen.

Chickens prefer grain to their layers' feed, so keep it for an afternoon treat. Whole wheat is ideal as a regular feed, but mixed corn is the chickens' favourite, especially the maize it contains. This makes it a useful aid to taming, and it's a warming feed for a winter's night – but in hot weather its 'heating' qualities may lead to aggression amongst the flock. Some of the grains in mixed corn may be too large for tiny bantams to manage easily.

Adding Supplements

Supplements can provide a useful boost when chickens are challenged by stress or sickness. Chickens living in a pen may need some extra support too.

There are a variety of supplements available, but apple cider vinegar (ACV) is widely used. Research has shown that it has many benefits, including bolstering the immune and digestive systems. It also helps chickens cope with the effects of stress (which can weaken their immune systems), and it contains several vitamins and minerals.

Make sure you buy unpasteurized ACV from a feed store or health shop – the type sold for cooking is pasteurized, which kills the friendly bacteria.

Apple cider vinegar should be added to the drinking water (20ml vinegar to a litre of water). You can either use it when you feel the birds need a tonic or give it regularly – one week in every month. It should only be used in plastic drinkers.

Giving Treats and Extras

Many people like to give their chickens some titbits, especially if they are kept as pets. Mixed corn is much appreciated, but there are a few other treats you can give without upsetting either DEFRA or the chickens' digestions.

New chicken keepers are sometimes shocked when they find their chickens busily dismembering a mouse or frog. Chickens eat meat as well as vegetable matter, and when free-ranging will pick up a variety of grubs. They are attracted to blood and love meat, but must not be fed it. However, you can give them worms and insects dug up from the garden.

Mealworms were once popular treats, but as they could have been raised on products of animal origin, are currently only permitted to be sold as wild bird food. The same goes for dried insects.

Pecking blocks of seeds are a useful diversion for confined chickens, but avoid any containing molasses if you have breeds with feathered heads and faces. Dried sunflower heads will go down well, and can sometimes be bought in feed stores.

What not to give

Don't give your chickens avocado, citrus fruit, dry pulses, uncooked potatoes or green beans, chocolate or anything salty.

Be careful not to overdo the treats – apart from piling on the pounds, an overload of goodies may cause digestive upsets.

Checking the Drinking Water

Always make sure your chickens have plenty of fresh water.

As mentioned, a laying hen can drink 500ml of water a day – even more in warm weather. All animals need water, but chickens have another ace under their wings to remind you. If left without a drink, they can stop laying for several days.

Check drinkers frequently on hot days and make sure they don't freeze during winter.

Even so, despite your best efforts to supply them with fresh water, chickens will happily drink from muddy puddles and any other unsuitable places they can find. This may cause disease, so keep the run as dry as possible, and try to fence free-range chickens away from areas of standing water.

Feeding Chickens

Left to their own devices, chickens spend much of the day looking for food – their diet is regulated by the need to work for everything they consume. If confined with little to do but eat a ready supply of food, they can become fat, which affects their health and laying performance. Boredom can lead to feather-pecking and other unpleasant habits.

Keeping your chickens in shape doesn't necessarily mean reducing their feed ration, as this can also have repercussions, but it makes sense to restrict fattening treats and grains when chickens are kept in a run. Green vegetables are a healthier alternative, and hanging them so the chickens have to stretch or even jump a little will provide a work-out too.

How much to feed

As a very rough guide, an average chicken will eat about 120 to 150g of feed a day, but this varies considerably depending on the size of the chickens, their breed and how they are kept. Active free-rangers will forage for much of their food, while confined chickens are dependent on what they are given. Weather also affects feed consumption – chickens will eat more in the winter, less if it is hot.

If you feed too little, those at the bottom of the pecking order will go without, so it's better to be generous at the beginning and provide more than enough for everyone. Ideally there should be just a small amount left over at the end of the day – you'll soon see how much is required to keep your chickens well fed and happy.

Fill the feeder in the morning and leave the chickens to help themselves. If you choose to scatter some grain in the afternoon, this will supplement their diet, but isn't essential. Take away any uneaten feed at night (chickens don't eat or drink once they have gone to bed).

Feeding new chickens

Like most animals chickens don't appreciate changes to their diet, and new feeds should be introduced gradually. When you buy chickens, ask what they are currently eating, and try to continue with this while they are settling in. If you want to make changes, replace a little of the current feed with the new one, increasing the proportions of new to old over the course of a week.

Rescue hens will require dry mash or ex-battery crumbs when first re-homed, although they can be gradually introduced to pellets later.

Young hens (pullets) shouldn't be fed layers' feeds until they are mature enough to produce eggs. The breeder should advise on age and feeding, but if in doubt continue with a growers' mix until they are ready to start laying. Giving layers' feeds too early can cause pullets to start producing eggs before their bodies are ready.

Avoiding Poisonous Plants

Chickens usually avoid poisonous plants, but you shouldn't rely on this, especially if your chickens live mainly in a run.

Free-range chickens with plenty of grass may take less interest in other vegetation. Ours have occasionally tucked into a plant, but will actually turn up their beaks when offered thinnings from the vegetable plot. Stealing is probably more fun.

Confined chickens are more likely to have a go at any greenery that comes within reach, and may be less able to discriminate between good and bad. Make sure that anything growing close to the run is safe for chickens to eat.

When your chickens are free-ranging in the garden, keep an eye on what they are nibbling. Remove them from anything you are not sure about and then check up on it.

Unfortunately there's limited conclusive evidence about how toxic some plants are to chickens. While some are definitely poisonous, others would probably have to be eaten in large quantities to do any damage.

This list isn't exhaustive, but your chickens probably shouldn't eat these:

Autumn crocus	Conkers	Henbane
Beans	Corn cockle	Holly berries
Bracken	Daffodil	Honeysuckle
Bryony	(especially bulbs)	Hyacinth
Buttercup	Delphinium	Hydrangea
Castor beans	Foxglove	Iris
Clematis	Hemlock	Ivy

Jasmine	Nicotiana	Rhubarb leaves
Laburnum seeds	Nightshade	St John's wort
Laurel	Oleander	Sweet peas
Lilies	Potato sprouts	Tomato leaves
Lily of the valley	Privet	Tulips
Lupins	Ragwort	Vetch
Mistletoe	Rapeseed	Wisteria
Monkshood	Rhododendron	Yew

Avoid using chemical pesticides or herbicides in areas where chickens are ranging.

Key Points

- Feeding kitchen waste to chickens is illegal, and can lead to an unbalanced diet
- Check expiry dates when buying feed
- Pellets are easy to handle and provide a complete diet
- Dry mash keeps chickens occupied longer but can be messy
- Mash can be fed moistened, but goes off quickly
- Chickens need insoluble (flint) grit to process their food
- Soluble grit provides extra calcium for strong eggshells
- Hang up some vegetables in the chicken run
- Only feed grain in the afternoon
- Unpasteurized apple cider vinegar has many health benefits – add 20ml per litre of water in a plastic drinker
- Don't feed avocado, citrus fruit, dry pulses, uncooked potatoes or green beans, chocolate or anything salty
- Too many treats can lead to health problems
- It's vital that chickens always have plenty of fresh water
- Feed generously to start with so that no chickens go without
- Make any changes to diet very gradually
- Don't feed layers' mixes to pullets until they are mature enough to produce eggs
- Chickens usually avoid toxic plants – but you shouldn't rely on this
- Don't use pesticides or herbicides where chickens are ranging

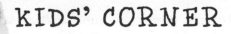

KIDS' CORNER

Quiz

What did you discover about feeding chickens in Chapter 6?

QUESTION ONE

What is the best food for chickens?
- (a) Ready-made chicken feed
- (b) Left-overs from the dinner table
- (c) Breadcrumbs

QUESTION TWO

Why do chickens need to eat little stones or grit?
- (a) It keeps their beaks sharp
- (b) To help break up their food
- (c) To make their egg-yolks yellow

QUESTION THREE

When would you feed chickens grain?
- (a) In the morning
- (b) In the afternoon
- (c) All through the day

QUESTION FOUR

Which of these items is bad for chickens?
- (a) Sunflower seeds
- (b) Chocolate
- (c) Apple cider vinegar

QUESTION FIVE

Which of these statements is wrong?
- (a) Chickens take longer to eat dry mash
- (b) Chickens like to eat grass or green vegetables
- (c) Chickens eat more in hot weather

ANSWERS

One (a); Two (b); Three (b); Four (b); Five (c)

Make sure you know how to feed chickens properly – in the next chapter we are going chicken shopping!

Chicken Chat

'Chicken feed': As you have just read, chicken feed comes in little pieces. This saying is used when talking about a very small amount of money or something that is hardly worth bothering about.

Chicken Jokes

Did you hear about the hens that went on strike?
They were fed up with working for chicken feed!

What do you get if you feed gunpowder to a chicken?
An eggsplosion!

Something to Do . . .

Where will you buy your chicken feed? Is there a store nearby, or will you have to order it to be delivered?

Find out which brands are most easily available, and have a look at the manufacturers' websites to see what will suit the chickens you are thinking of buying. Write down the items you will need – pellets, mash or crumbs; grain; flint grit; soluble grit; apple cider vinegar; supplements (remember that at first you should give your new chickens the feed they have been used to, and make any changes gradually).

CHAPTER SEVEN

How Do You Cock-a-Doodle-Doo? Buying Your Chickens and Introducing Them to Their New Home

The chicken house is set up in the garden, and everything is ready. All that's needed now are some chickens.

Deciding How Many to Buy

A solitary hen will be miserable, so two is the minimum – although it's better to have three in case you lose one.

Even if you are planning a sizeable flock, starting with a manageable number gives you a chance to gain experience before you expand. Four to six hens would be ample. After a year or so you can add new hens, which will boost the egg supplies as the older girls begin to lay less.

As mentioned earlier, it's really important not to overstock.

Mixing Different Breeds

A mixture of breeds can look attractive, particularly free-ranging around the garden, or you could combine the appeal of a few pure-breeds with the productivity of hybrids.

Some breeds are liable to be bullied in a mixed flock, so choose birds that can hold their own – especially if they are kept in a run. Keep sizes similar too, although some bantams can be bullies. When introduced to her new Cochin room-mate, my little Pekin Bantam immediately established who was boss – even though she had to leap a couple of feet into the air to do so.

Choosing to Buy Chickens, Chicks or Eggs

Chickens can be bought at all ages from the egg upwards.

Point of lay

It's easiest to start with young hens at what is called 'point of lay' (POL). This term tends to be used loosely and thus causes a fair amount of confusion.

New chicken keepers often (not unreasonably) expect that a point-of-lay pullet is either laying or about to start. A more accurate definition would be that she is approaching the age when she will be mature enough to produce eggs.

Actually, it's better if the hen hasn't begun laying – a house move at this delicate time of her life can upset her and cause problems.

Point-of-lay pullets may not produce eggs for a few weeks or even longer. It's disheartening to look at an empty nest-box every day, but different breeds begin laying at various ages. Hybrids may lay their first egg before they are twenty weeks old, while it could be at least another month or two before some of the pure-breeds are mature enough to lay.

The time of year also makes a difference. Pure-breeds who reach maturity in the autumn often don't start laying until the following spring. Hybrids are more likely to start as soon as they are ready.

When buying point-of-lay hens, ask their actual age and when they are likely to start laying. This will also ensure you know which feed to give them.

Older hens

Sometimes you might be offered older hens. They may not provide as many eggs as hens in their first year, but should be cheaper to buy. Hens more than three or four years old will be past their laying best.

If you want some ex-battery hens, you should contact one of the re-homing organizations.

Eggs and chicks

It may sound appealing, but there are several drawbacks to buying your first chickens as hatching eggs:

- Even the experts have failures when hatching eggs in an incubator
- The eggs will contain an unknown mixture of both males and females
- Incubators can be expensive, especially the models that do much of the work for you
- You will need a 'brooder' to keep the chicks warm until their feathers grow

Day-old chicks are also available. Even if they have been sexed, mistakes can be made, so there's no guarantee there won't be any males. Raising healthy birds from young chicks takes some experience, so this isn't an ideal way to buy your first chickens.

Chicken Shopping

Don't rush when buying chickens. A hasty purchase may come back to peck you!

Purchasing from a breeder is usually the safest option. A reputable breeder should be happy to answer all your questions and will want to ensure that you are satisfied with your chickens.

Some suppliers of poultry housing also sell chickens – and some chicken breeders can supply coops. You may be offered a complete package of hens, house and accessories, but look carefully at what is included and check that it's suitable for your requirements. Make sure the house is up to standard, equipment is good-quality and that you are getting healthy hens from a reliable source – it might not be the great deal it seems.

A few garden centres and pet shops now sell chickens, as well as housing and supplies. Quality can be variable, and the assistants may have little experience of chickens. Buying chickens from a shop is often more expensive than from a breeder, choice will be limited and there may be little information about the birds themselves.

There are also auctions of poultry, which are exciting but can be risky.

Buying from a breeder

Chicken breeders advertise in poultry magazines (see Further Reference), and most of the popular breeds and hybrids are offered. Many of the larger breeders have websites.

Ask around too – if you can find someone through personal recommendation so much the better.

Try to find a suitable breeder close to home. Apart from being more convenient, this saves subjecting your new chickens to a long and stressful journey. It will also be easier if it's necessary to return any birds.

Some breeders are enthusiasts specializing in only a few pure-breeds. They will often have won awards at shows and may have surplus stock for sale – usually birds unsuitable for showing but fine as domestic chickens. If you want a particular breed, this is a good way of obtaining chickens from an expert. Specialist breeders can be tracked down through the relevant breed society.

Larger breeders will offer a wider selection. Some breeders sell both pure-breeds and hybrids, while some concentrate on one or the other.

Hybrid hens are usually available all year round, while spring/summer is a good time to start looking for pure-breeds. Breeders can only raise so many birds at one time, so if you have set your heart on a popular breed, you may have to wait until they are ready.

Looking around

Try to visit one or two breeders before you are ready to buy your chickens. You can have a look around, discuss your requirements and put your name on a waiting list if necessary. Some breeders will expect you to make an appointment before visiting, but should be happy to show you their stock and take an order for the future. Going without the intention to buy also makes it easier to walk away if you are unhappy about anything.

You should find the birds being kept in clean, spacious conditions. Overcrowding in a dirty, damp environment is the cause of diseases in young chickens, especially coccidiosis, a parasitic disease of the intestines which often results in death.

The chickens should be bright and alert. Are any sitting hunched in a corner? This is a sign of bad health and poor husbandry too, as they should have been removed. The chickens' eyes, nostrils and rear ends should be clean with no discharge. Feathers should be bright, shiny and intact.

Ask whether the young birds have been vaccinated, and against which diseases. These might include: coccidiosis, infectious bronchitis, Marek's disease, mycoplasma, Newcastle disease and salmonella. Most vaccinations can only be given at a young age and shouldn't need repeating.

Ask if you can see the parent birds. They also should be bright and healthy. How do they react to their keeper? Ask about flightiness and ease of taming.

The breeder should be happy to give you help and advice, as well as being prepared to exchange any unsatisfactory birds.

A word of warning

There are some unscrupulous breeders motivated more by profit than providing good-quality birds – the chicken equivalent of puppy farmers. Cutting corners to make quick sales usually results in unhealthy birds likely to succumb to disease. These breeders may be able to supply exactly what you want when you want it, but will have no interest in dealing with any problems that arise as a result of their shoddy practices. They thrive on people who are too impatient to wait until the reliable breeders have re-stocked – you have been warned!

Buying privately

You might find chickens advertised for sale privately – perhaps on a board in the feed store. These adverts may be placed by small-scale breeders or people who have just hatched out a few eggs from their own hens. Quality may vary, so look at the birds and premises carefully, making sure you ask plenty of questions. Hybrid hens are usually sold through agents, so be cautious if someone is advertising 'Black Rocks' or other well-known names. They could simply be cross-breeds and, although they may look similar, will not have the qualities expected from hybrids.

Buying from sales or auctions

The normal advice to the novice is never to buy from sales, but they are interesting places to visit, if only to look at the different breeds. There may also be chicken equipment available (thoroughly disinfect anything bought second-hand), trade stands and a chance to mingle with other chicken enthusiasts.

Salisbury Poultry Auction is one of the largest in the country, and close enough to provide me with an enjoyable morning out. Friends have had some successful buys, and I have some hens still going strong after a couple of years – as well as a cockerel who started crowing a few weeks after he was auctioned as a year-old hen . . .

That's the problem with sales – even if the mistake is the seller's, there's no redress. If you buy a sick bird, a hen that turns out to be a cockerel or something listed incorrectly, you are stuck with it. As chickens become ever more popular, it's likely that more and more unsuitable birds will end up at sales to be bid for enthusiastically by the unwary.

It's all too easy to get carried away by the excitement of an auction, and I've seen people pay well over the odds. Sales really aren't the best way of buying your first chickens – or even your second.

Avoiding common pitfalls

If you are still keen to buy at auction, here are a few tips:

- Sales are noisy, confusing places, so first go and have a look round without intending to buy.
- Try to obtain a catalogue before the sale, and mark any lots that interest you – Salisbury Poultry Auction publishes its catalogue online.
- Obtain a buyer's number before bidding – the auctioneer will need this if your bid is successful.
- Arrive early so you can study your intended purchases before bidding

starts. There is often a crush around the auctioneer, making it difficult to see the birds.

- Look carefully at the chickens you intend to buy, and reject any that don't appear healthy.
- If a bird looks unwell, don't consider others from the same seller.
- Sometimes lots are changed or aren't specified in the catalogue. Make sure you know what the breeds should look like.
- Chickens are often sold as 'pairs', 'trios', 'quartets' or 'quintets'. All of these contain a cockerel. There should also be some lots consisting only of pullets or hens. 'As hatched' means the birds haven't yet been sexed.
- There will be a buyer's premium added to your bid and VAT if the seller is registered. Decide how much you are prepared to pay and stick to it.
- You will need a pet carrier or box for your new chickens – you may be able to buy cardboard pet boxes at the sale.
- Take your chickens straight home and give them quiet, comfortable quarters with plenty of food and water.
- Quarantine chickens bought from sales for at least three weeks.

This last point is particularly important if you already have some hens and don't want to spread potential sickness. Sales are a great place for infections, and stress makes chickens particularly susceptible. Birds bought from different sellers should ideally be quarantined separately.

Bringing Your Chickens Home

Proper carrying boxes are costly and must be disinfected after every use. For transporting chickens from breeder to home, strong cardboard boxes should be adequate. The box should be large enough for a chicken to stand up and turn around, and you will need enough boxes to carry all your purchases comfortably – preferably one box per chicken. Don't forget to make plenty of air holes. Make sure the top can be closed securely and take some string so you can tie it shut.

Cardboard pet carriers can be bought cheaply – sometimes the breeder will supply these, but check first so you don't have to go home empty-handed.

You could also use a cat carrier, or even a small dog crate – cover it with a blanket so the chickens aren't panicked by the journey.

Line your boxes with newspaper and add a layer of shavings.

Chickens overheat very easily, leading to stress or even death, so keeping them cool and comfortable is a priority when travelling.

Don't put chickens in the boot, as it may become too hot and airless. Putting them in the car allows you to monitor the temperature – chickens should never be left in a parked car on a warm day even with the windows open. Try not to transport chickens on very hot days – if this is unavoidable, travel early in the morning or late in the evening. Cardboard boxes generate warmth, so use an alternative for long journeys in warm weather.

Pet carriers can be used for transporting chickens

Food and water must be provided for journeys over eight hours – but hopefully you won't need to travel that far.

Settling Your Chickens in Their New Home

Before going to collect your chickens make sure everything is ready for them, so they can be put into their new quarters with the minimum of delay or fuss.

Your henhouse should be set up and the run prepared. Spread a layer of bedding in the henhouse. There should be enough to comfortably cover the floor, but it needn't be very deep as it's mainly there to absorb droppings – be more generous if the weather is particularly cold or wet. Ex-battery hens need a thicker bed to give them protection and warmth, especially as they may sleep on the floor at first.

Provide a good layer of bedding material in the nest-boxes. If you have bought some diatomaceous earth, sprinkle it around the house and nest-boxes.

Fill the feeder, drinker and grit hopper.

The chickens may be stressed after the journey and the change to their surroundings. They have probably been kept with many other birds, and now find themselves on their own. It may take a little while before they feel confident in their new home, but they will settle more quickly if their re-location goes smoothly and is as stress free as possible.

Transferring the chickens

You may feel rather nervous if you haven't handled chickens before, but try to keep this process as calm and relaxed as possible. Shut other pets away, and ensure that everyone stays quiet while you move the chickens from their carrying boxes to their new home.

Open the box enough to slip your hand in and locate the chicken. Talk quietly to the bird – it doesn't matter what you say, but keep your tone gentle. Press lightly on the chicken's back with one hand, bringing your fingers round to keep the wings from flapping. Slide your other hand under the bird, so she is sitting on your palm with her legs between your fingers. You should now be able to lift her gently out of the box, and place her in the house.

Handle your new chickens with confidence (even if you don't feel it!), and hold them securely but not too tightly. Stopping the wings from flapping is important – once you have lifted the hen from the box, you can tuck her under your arm while her front stays comfortably balanced on your hand. Keep the rear end pointing away from you to avoid messy accidents! Don't ever pick a chicken up by the legs or grab a wing.

Alternatively you could place the carrying boxes in the house and allow the hens to make their own way out. Although you'll have to handle the birds at some point, this may be easier for now and cuts down the risk of chickens escaping.

Putting the new chickens straight into their house will help them to realize where home is and give them a feeling of security. Leave them shut in for an hour or so before letting them out into the run – put a drinker in their house if they have had a long journey. If they have arrived in the late afternoon, offer some food and water in the henhouse, and then leave them and let them settle down for the night.

If the chickens are to be free-range, either rig-up a temporary run or provide food and water in the henhouse and keep them shut in for a couple of days.

Settling in

It will be very exciting finally to have the chickens after all the preparations, but try to keep their surroundings peaceful – no barking dogs or noisy children. Talk to them when you approach, and keep movements slow and deliberate so as not to startle them.

Having been careful not to let your chickens out while they were acclimatizing, you may be surprised not to be knocked down in the rush when the pop-hole is finally opened. Sometimes it takes them a while to summon up the courage to venture out, but if you put a feeder where they can see it and scatter some corn, they will eventually emerge.

Time for bed

Once they have come out, they should find their own way back at night, but don't count on this. Ex-battery hens especially can take a little while to understand the difference between night and day.

If you find your chickens have settled down in a corner of the run, pick them up gently (this will be easy when they are drowsy) and place them on their perches. Ex-battery hens should be put on the floor of the house, as they may be too weak to jump down safely from perches – wait until they are strong enough to hop up by themselves.

Make sure free-range hens are moving towards their house as evening approaches (put the feeder by the pop-hole), or you may end up searching for them with a torch!

Chickens will look for a safe roost for the night, and usually return to their henhouse once they have spent some time there. Soon they will happily put themselves to bed – you will only need to check they're in and close the door.

Establishing a pecking order

The chickens will quickly begin sorting out their social structure.

What feathers ex-battery hens have left may start to fly – they are adult birds and unused to living together.

Pullets will also need to arrange their hierarchy. Sometimes peace can break out after a few pecks, but if two hens want the top role you could find yourself witnessing a 'hen fight'.

These battles can be distressing, but they have to be fought, so try not to interfere physically unless necessary. The following tactics should help:

- Extra feeders and drinkers can help avoid squabbles
- Hang up some vegetables to provide a distraction
- String up some old CDs – chickens are attracted to shiny objects
- Logs, branches and small bushes will provide entertainment, as well as hiding places
- Clapping your hands or shouting 'Oi!' (as if to a group of naughty children) may distract the hens – and relieve your feelings
- Chickens go for their adversaries' combs, which can bleed quite alarmingly, although they soon heal – petroleum jelly on the combs prevents hens from getting a grip
- Allowing confined birds out to free-range for a while can calm things down

Free-range hens usually settle more quickly than enclosed birds. They may still fight, but there is more space, and lower-ranking hens can easily hide.

Bullying

Make sure that no hen is being badly bullied. Blood is a magnet to chickens, so quickly remove any birds that are bleeding to avoid further attacks.

Occasionally a hen keeps bullying the lower ranks – or one bird is picked on by the others. If a hen is having a particularly hard time, she may be prevented from eating. Removing her completely will only make matters worse when she eventually re-joins the flock, but you could try giving her food and water just outside the run. This will ensure she doesn't starve and help build her confidence. Stay close when she is feeding with the others again, reassuring her with your presence.

It's usually better to remove the bully rather than the victim. Put her in a separate pen where she can see the others but not attack them. Being taken out of the flock for the day may reduce her status and give the others a rest too. Put down several food and water containers when she returns to avoid further confrontation.

Chickens generally settle down easily, and once they have sorted themselves out the flock becomes a strong unit, with each one knowing her place.

Taming

If your chickens are tame enough to be picked up easily, it's much simpler to examine them or to administer treatment.

Talk regularly to your chickens – they will soon come to recognize your voice and know you as the person who brings food.

Chickens may enjoy human company, but food is what really counts. Offer some corn and most chickens will want to be your friend. Some might immediately take treats from your hand, although others will be more suspicious.

Once they are confident with you, progress to gentle stroking – chickens are easily spooked by overhead objects, so bring your hand from below rather than above as you reach to stroke their back.

With your chickens running to take treats from your hand and submitting to being stroked, you can proceed to picking them up. Never grab or move quickly. Slide your hand under the breast, fingers spread to take the legs, so the chicken is balanced on your palm. Support the back and wings with your other hand. Put the chicken down after a short while and reward it with another treat.

You may find the feel of feathers strange, but this usually passes as you become used to handling chickens.

Taming will be harder with some of the flightier breeds, and some never enjoy being handled. Sometimes it's better for free-range birds to remain slightly cautious so they are alert to danger.

Pick your times for chicken taming. They are less likely to be interested when they're full for the night and ready to go to roost. When initially let out, they will be anxious to be first at the food and water, and may ignore your overtures of friendship.

You can train your chickens to come when you call by associating treats with a certain sound. Try rattling some corn in a tin – they will soon cotton on to this and come running at the sight of their tin. This is particularly useful if you need to get them back into their run quickly.

Should You Name Your Chickens?

It's fun to christen pet chickens, although it's unlikely they will ever learn their names.

Sometimes their appearance or personalities provide inspiration – we had Lara Croft, who marched indoors and stormed the stairs. What about their country of origin? Our Cochin hen is Lotus Blossom (Cochins come from China). A friend named her chickens after major political figures.

If you intend to name your chickens, choose different breeds or at least different colours so you can easily tell them apart.

Adding to Your Flock

Eventually you may want more chickens or need to replace losses. While settling a new group of chickens can result in battles, adding to an established flock can start a war! The flock will see new chickens as intruders, and are likely to attack them. Some breeds are more accepting than others, but don't add just one or two hens to a group of several – the newcomers will have a very hard time and may even be killed. Ideally the new hens should equal or outnumber the existing flock.

If you fancy some new breeds, consider whether they will mix with the hens you already have. Adding an easily-dominated type to a flock of assertive chickens is a recipe for disaster.

Before buying more hens, you should acquire a spare house and run so the new arrivals can be quarantined. Additional housing will always be useful for any birds that need isolation.

Position the housing so the hens can see each other, but are not close enough to pass on any disease (make sure you don't transfer infection yourself via clothing or equipment). Take this opportunity to worm all the hens and treat for parasites.

The new birds should be kept separately for about three weeks – don't introduce them to the others until you are sure that all is well.

Making the introductions

This is best done at night. Once all the chickens have settled down, pop the new girls on to the perches amongst the old ones.

All will be well while it's dark – cover any windows so that the birds stay in darkness until you let them out.

There will be scuffles when you do, but plenty of feeders, drinkers and distractions will help. Confined hens will need longer than free-range birds to sort out their differences, but as long as they aren't overcrowded they should settle down eventually.

Key Points

- Two is the minimum number of hens to keep – three is better
- Even if you plan to expand, it's easier to start with a manageable number
- Never overstock
- Don't mix timid breeds with assertive ones
- Start with pullets at point of lay
- Birds bred in poor conditions may have diseases
- Chickens for sale should be alert with good feathering, clean eyes, nostrils and bottoms
- Ask about vaccinations
- A good breeder will give advice and be willing to exchange unsatisfactory birds
- Auctions are interesting to visit, but full of pitfalls when buying
- Keep chickens cool and comfortable when travelling
- Prepare everything before the chickens arrive, put them straight into their house and keep their surroundings peaceful
- Hens must establish a pecking order, so provide extra feeders and distractions
- Separate bullies and victims, but only if necessary
- Food encourages chickens to become tame
- A flock will see new chickens as a threat
- New chickens should be quarantined for at least three weeks

KIDS' CORNER

Quiz

Are you ready to buy some chickens now?

QUESTION ONE
What would be a good number of chickens to start with?
- (a) One
- (b) Three
- (c) Twenty

QUESTION TWO
What is the easiest way to buy your first chickens?
- (a) As hatching eggs
- (b) As day-old chicks
- (c) As point-of-lay hens

QUESTION THREE
Where would you be most likely to buy some satisfactory chickens?
- (a) From a well-known breeder
- (b) At an auction or market
- (c) From a pet shop

QUESTION FOUR
How will you bring your new chickens home?
- (a) In a travelling box on the back seat of the car
- (b) On your lap
- (c) In the car boot

QUESTION FIVE
How should you pick up a chicken?
- (a) By the legs
- (b) By the wings
- (c) With one hand underneath and one on top

ANSWERS
One (b); Two (c); Three (a); Four (a); Five (c)

Did you get all the answers right? Check Chapter 7 again before buying your first chickens!

Chicken Chat

'The chickens have come home to roost': This is used when someone faces the consequences of a foolish or bad action. For example: 'Sam was always teasing the girls, but his chickens came home to roost when Natalie stood up to him'.

Chicken Jokes

What is the best time to buy chickens?
When they are going 'cheep'!

Did you hear about the farmer who bred three-legged chickens?
They would be worth a lot of money – if only he could catch them!

If fruit comes from a fruit tree, where does a chicken come from?
A poul-tree!

Something to Do . . .

Make friends with your new chickens. Keep talking to them quietly and they will soon start to know your voice. Offer them some treats to encourage them to come closer. Always move slowly and don't do anything to scare them. Before long you will be able to tell where each one is in the pecking order, and even if they all look alike you will begin to recognize them by their different characters.

CHAPTER EIGHT

Doing the Hen-Housework: Caring for Your Chickens

There are as many ways of keeping chickens as there are chicken keepers, and you will soon work out what suits you and your flock. Chickens living in a run usually require more attention than free-range birds.

The following suggestions should be adapted to individual circumstances.

Establishing a Daily Routine

Everybody's day is different, but this is the basic minimum you will need to do.

Morning
- Put out full feeders and drinkers, top-up the grit hopper
- Let out your chickens: 'Good morning, chickens'
- Watch the chickens come out – they should be keen to eat and drink
- Check for eggs in the house and nest-box
- Make sure fencing is secure (test electric fence)
- Hang up some vegetables in the chicken run

Chickens don't have to be let out at first light – this could be four o'clock in summer, and predators are often around in the early dawn. They will wait until a reasonable hour if their coop is kept dark, but don't leave them too long as they need daylight for egg production.

Visual checks
As the chickens come out of their house, keep an eye open for any of the following potential signs of illness:

- Not eating
- Not drinking or drinking excessively
- Lethargy, closed eyes
- Bleeding or wounds – remove the bird at once
- Limping or a wing trailing
- Dull, ruffled feathers or feather loss (apart from when moulting)
- Pale or purple comb (combs shrink and go pale when moulting)
- Discharge from eyes or nostrils, coughing, sneezing or rasping
- Diarrhoea
- Feather pecking or bullying

Chickens don't usually look ill until something is really wrong. Any bird that appears hunched or depressed should be removed from the others and thoroughly checked.

If you spend time watching your chickens, you will soon notice any variations to their normal behaviour and develop a feeling for when something is wrong.

Listen to Your Chickens

Research has shown that chickens make around thirty different sounds. Some of these are made only by cockerels or chicks, but you can still pick up a variety of communications amongst a flock of hens.

There are cries of triumph when an egg is laid, shrieks of rage when the neighbour's cat approaches, and different alarm calls depending on whether a predator is air- or land-based. Your hens will make contented noises when they are happy and angry ones if something is wrong.

Learn what your chickens are saying, so you will know whether to rush out with a broom or put the pan on for a new-laid egg!

Evening
- Collect the eggs (remove any droppings from nest-boxes)
- Remove the feeder and drinker – clean them if necessary
- Store reusable feed in a vermin-proof container
- Clear any food or vegetables from the run
- Check that all the chickens are in and settled: 'Goodnight, chickens'
- Securely close all pop-holes, doors and nest-box lids

Collecting eggs

Collect eggs regularly to discourage egg-eating or hens going broody. Pullets often lay outside the nest-boxes, so check for eggs in the house and run.

An unexpected drop in egg production or poor shell quality sometimes indicates a problem (see Chapters 10 and 11).

Removing the feeder and drinker

Don't be tempted to leave the feeder and drinker in the henhouse overnight in case the chickens fancy a midnight snack. Chickens don't eat or drink once they've gone to roost, but the water will make the house damp and the feed may attract rodents.

Other daily tasks

You might also need to:
- Check drinkers more frequently in hot or freezing weather
- Pick up droppings from the run and henhouse – keep a bucket and shovel (or rubber gloves) handy
- Add disinfectant/sanitizer to the run
- Shift housing to fresh ground
- Scatter an afternoon feed of grain
- Let confined chickens out for some free-ranging
- Enjoy quality time with your chickens, encouraging them to become tame

Doing the Weekly Chores

Some jobs can be done weekly:
- Cleaning out the henhouse
- Cleaning the run and dust-bath
- Repairs and maintenance
- Giving the hens a thorough physical check

Cleaning out the henhouse

Probably not a favourite job, but an important one – remember that chickens do most of their droppings at night.

Arm yourself with rubber gloves and wear old clothes or overalls. Keep your cleaning-out tools together where they are easily accessible.

Take out the perches and droppings board. Remove the dirty bedding to the compost heap – make sure the chickens can't play with it.

Clean out the nest-boxes – you can save on bedding by recycling clean nest-box material to use on the henhouse floor.

Use a scraper and stiff brush to give the house, perches and droppings board a thorough clean.

Leave the house to air with the door open in fine weather. Disinfectant powder helps dry up wet patches.

Put in fresh bedding – preferably an hour or so before the chickens go to roost so any dust can settle. Add insect repellent or diatomaceous earth as necessary.

When cleaning out, look for any signs of red mite (see Chapter 11) and take immediate action. They multiply very quickly, especially in warm weather.

Plastic houses can be hosed out – as long as the house is thoroughly dry before the chickens go to bed.

Give the feeder and drinker a good scrub as part of your cleaning routine – disinfect them occasionally.

Cleaning the run and dust-bath

If you regularly pick up droppings in the run, the weekly clean will be much easier. Remove droppings and any spilt food. You may need to turn the run material over, change it or hose it down. Disinfect and sanitize regularly, especially in warm weather.

Remove feathers and droppings from the dust-bath. Add fresh sand as necessary. Diatomaceous earth can be sprinkled into the dust-bath to help with parasite control.

Repairs and maintenance

When cleaning the henhouse, see if anything needs attention. Find loose door catches or weak spots in the run before the fox does! Look for any evidence of rats – droppings, wood being gnawed, small tunnels – and deal with them quickly (see Chapter 9).

Check electric fencing and trim any vegetation that may touch it and cause it to lose power.

Giving the hens a thorough physical check

If the chickens are unwilling to be caught, wait until evening when they can be easily picked off their perches. A large landing net is useful for catching a chicken in the daytime – with a little practice it can be dropped over the bird fairly easily. Don't try chasing chickens. You are unlikely to catch them but will cause them stress (and end up hot and stressed yourself too).

Hold the chicken securely but not tightly – enlist an assistant if possible –

and check the following (see also Chapter 11):

- Are there any injuries or swellings?
- Does the hen feel lighter or heavier than usual? A sharp breastbone is a sign of weight loss.
- Look carefully under the feathers for any signs of lice or mites, and dust the bird with mite powder when necessary.
- The crop (at the base of the neck) will be full if the chicken has just eaten, but should be loose and empty by the morning.
- Eyes, ears, nose and vent should be clean with no discharge.
- Breathing – are there any rattling noises?
- Legs and feet – scales should be flat, with no build-up of debris on feet.
- If nails are starting to curl under, trim the tips with nail clippers, cutting below the dark line – a styptic pencil is useful, just in case the nail bleeds.

Remembering Other Tasks

There will be other occasional jobs to do. These might include:

- Worming
- Changing the run material
- Moving electric fencing
- Scrubbing the henhouse
- Preparing the henhouse for winter
- Wing clipping

Worming

All chickens are at risk from several kinds of internal worms. They are passed on through droppings or may be transmitted via wild birds, earthworms, slugs and snails.

How worms affect chickens

A chicken with a worm problem will become thin and listless. The feathers will look dull and rough, and the comb pale. The chicken may have diarrhoea and worms are sometimes visible in droppings. Egg production will be affected, with paler yolks – worms have even been known to appear in the eggs! The bird will be weakened and susceptible to disease or may die from the build-up of internal parasites.

Worming a chicken with a heavy infestation can also result in death due to the toxins released from the worms when they are destroyed. Regular worming helps to keep internal parasites from reaching dangerous levels.

Using wormer

Flubenvet wormer is licensed (and therefore fully tested) for chickens, and can be purchased from the vet, via online outlets and from some feed stores. It should only be sold by a suitably qualified person (SQP). When buying online you should be asked to complete a form and be given additional information about the drug.

Flubenvet is added to feed over a seven-day period. Mixing the powder with a little vegetable oil helps it to stick to feed pellets rather than sifting through to the bottom. It can also be mixed with wet mash. Alternatively you can buy pellets that include Flubenvet, but this might prove expensive if the feed passes its expiry date before it is time to worm again.

Your vet can prescribe other types of wormer, but there will be an 'egg-withdrawal period' when you cannot eat the eggs. This will be advised by the vet. Flubenvet has no withdrawal period, so you can continue to eat the eggs, although not the chicken itself until seven days after treatment has finished.

The general advice is to administer Flubenvet at least every six months – in early spring when mild days encourage worms to hatch, and again in autumn to clear up the summer crop. Winter weather should then stop any more emerging until the following spring.

Your chickens may need worming more frequently than this. Chickens in a static run are likely to be more challenged by worms than those regularly moved to fresh ground. Ex-battery hens will have no natural immunity to worms. Consult your vet for advice on individual worming regimes or if you suspect a bad infestation.

When dealing with an infestation of worms in your birds be prepared to dose again after a couple of weeks. This is to destroy the worms that will have hatched after the first treatment.

New stock should be wormed before being added to your flock.

Remember that Flubenvet is a drug – follow the instructions carefully, wear gloves and dispose of unused wormer properly.

Worm-egg counts

Worm levels can be checked by a worm-egg count. This involves sending a sample of droppings to a laboratory, either via your vet or through an online company (see Further Reference), who will post you a collection kit.

Natural prevention

Herbal products are available which help prevent and may reduce worms. You might prefer these as an alternative or back-up to Flubenvet. For example, you could dose with Flubenvet twice yearly and use a herbal product monthly. Or you could take a totally herbal approach and have regular worm-egg counts done to make sure all is well.

Herbal products won't kill gapeworms, which attach themselves to the bird's windpipe causing it to gasp for air (gape). Young birds are most often affected, especially if they are in contact with pheasants and other wild birds. Gapeworms aren't very common, but will cause fatalities and should be treated with a chemical wormer.

Apple cider vinegar is believed to make the chicken's interior less worm-friendly, as are crushed garlic cloves in the drinking water or Diatom added daily to the feed.

You can protect your chickens by keeping them in clean surroundings too. Ground used by chickens for long periods will become heavily infested with worm-eggs, which is why it's important to clear droppings regularly from the run and treat the ground with a product that destroys worm-eggs and larvae.

Free-range birds should be moved to new ground occasionally to allow the pasture to recover. Keeping grass short in summer allows ultraviolet rays to reach the worm-eggs and destroy them.

Changing the run material

Straw needs changing often, especially if it gets wet, while hardwood chips can last a few months.

Dig everything out and clear up droppings. A solid base can be hosed down with disinfectant. An earth base should be treated with ground sanitizer.

Moving electric fencing

Electric fencing should be moved occasionally – how often will depend on the size of the area and the number of birds. The job is easier with two people, and if the netting is rolled carefully it should be straightforward to re-erect.

Remember to trim vegetation around the new fence line.

Treat the used area by harrowing or raking and re-seed if necessary.

Scrubbing the henhouse

The henhouse should be scrubbed out occasionally. Start early on a sunny day.

Remove everything from the henhouse, take it apart if possible and scrape it thoroughly. Scrub with a detergent such as Poultry Shield, getting it into all the cracks. Allow to dry completely before adding fresh bedding.

An attack of red mite will necessitate extra cleaning, as explained in Chapter 11.

Give the feed storage bins an occasional scrub as well.

Preparing the henhouse for winter

Ensure the house is watertight before winter sets in. Check the roof, floor, doors, windows and ventilation holes. Deal with any renovations while the weather is reasonable. The alternative is finding a houseful of soggy chickens one bleak morning and struggling with emergency repairs in a snowstorm!

Wooden houses should be re-proofed as directed by the supplier. Make sure you use a product that is suitable for chickens (ring the manufacturer if in doubt), giving it plenty of time to dry and air.

You may decide to re-locate your chickens to winter quarters. As the grass disappears, a movable house and run could be placed on hardstanding, paving or a patio covered with hardwood chips. Another option is to adapt a large shed or outbuilding so the chickens can live indoors during the worst of the weather.

Wing clipping

Chickens kept mainly in a roofed run and heavy breeds that struggle to hop over a low fence shouldn't need their wings clipped. Free-range birds are better able to escape danger if they can fly.

However, some breeds are able fliers, and you may need to stop them from exploring neighbouring gardens or roosting in trees. If it's essential to clip your chickens' wings, here are some guidelines:

- Clipping doesn't hurt the chicken if it's done properly.
- Large, sharp scissors will be required.
- Try to recruit a helper – a wriggling chicken and scissors aren't a good combination. Alternatively, wrap the bird in a towel.
- Only one wing is clipped. This unbalances the bird and stops it taking off.
- Spread out the wing and you will see the ten long, pointed 'flight feathers', which extend from the elbow of the wing to the tip. Don't cut any other wing feathers.
- Check the feathers are fully grown – the quills should be white and hollow. Quills that are still growing are darke, due to the blood they contain, and must not be cut.
- If you accidentally cut into the blood supply, the bird will bleed copiously. Staunch the blood and seek immediate veterinary attention.

- Trim the flight feathers to about half their original length.
- Overlaying the flight feathers is a row of shorter feathers. You can use these as a guide, cutting below them to give a straight line.
- You will need to repeat the process every year when the feathers have grown back after moulting.

If you feel nervous about wing clipping, ask the breeder to show you when you buy your chickens or consult your vet.

Can chickens fly?

That was the question being debated in the bar by the officers of the British Army's 1st Airborne Division on a summer evening in 1944. The men were restless after a series of cancelled operations, and Lt Pat Glover decided to prove his point – that as chickens had wings and feathers they should be able to fly.

He kidnapped a chicken from a neighbouring farm and took 'Myrtle' with him on his next parachute jump. Once they had left the plane, he opened the canvas bag that contained the hen. Unsurprisingly, she took one look and decided to stay put. When they got down to fifty feet, Lt Glover released Myrtle and she managed to flap her way safely to earth.

Myrtle made many more jumps. Eventually she could fly from 300 feet, and would wait on the ground for Pat Glover to join her. She became known on the base as 'Myrtle the Parachick' and was awarded her parachute 'wings'.

When the 1st Airborne were finally called to action, Myrtle and Lt Glover were amongst the troops deployed to take a strategic bridge at Arnhem. They expected little enemy resistance, but Pat Glover kept Myrtle in her bag when they dropped into the Netherlands. He found himself descending into a raging battle, but still took care not to roll on Myrtle's bag as he landed. He handed the hen to his batman, Private Joe Scott, and joined the fierce fighting.

The following evening the paratroops were under attack from German machine guns and swiftly dug themselves into a trench. In the chaos Myrtle's bag was left on the edge and was found riddled with bullets. Like many others, Myrtle the chicken didn't survive the disastrous assault on Arnhem.

Pat Glover and Private Scott carefully buried Myrtle in her canvas bag, still wearing her parachute wings. Private Scott gave her a fitting tribute: 'She was game to the last, sir,' he said.

Arranging Cover for Holidays

It's essential to make sure your chickens are properly cared for when you go away – even if it's only for a couple of days.

If you have automated equipment, the chickens must still be checked at least once a day. The equipment may fail, a sick or wounded bird could be attacked by the others and eggs should be collected. Leaving chickens completely unattended could result in a charge of neglect.

Here are some ideas to make sure you have both a relaxed holiday and happy chickens:

- Ask a neighbour to help. This could be the obvious option as long as your neighbours are willing to take on the task, are reliable and understand what is involved.
- Take the chickens to family or friends. A small ark could be moved to another garden – remember there may be unfamiliar pets or predators in the new environment.
- Start a neighbourhood chicken watch. Get together with other chicken keepers and form a reciprocal sitting circle.
- Hire pet or house sitters. You could pay someone to look after your chickens or even to stay in your house while they do so. Look for adverts in feed stores, pet shops, the vet's surgery and country magazines, or try the internet. Make sure your pet sitter is reliable and knows how to look after chickens – you will be giving a stranger access to your home so check references carefully and arrange to meet in person beforehand too.
- Use a chicken boarding service. These are fairly widespread – try an internet search to see what is available in your locality. Check out the premises thoroughly, making sure security, maintenance and standards of cleanliness are high. Coops should be thoroughly disinfected between occupants and relocated to fresh ground. Take a look at the current guests too – do they seem happy and healthy?
- Take your chickens with you. This may be possible if you are in a suitable self-catering property, and can transport both chickens and the henhouse, but remember that long journeys in hot weather aren't good for chickens. Look out for unfamiliar predators at your holiday home – badgers, mink, pine martens and birds of prey.

Your chickens will usually be happier in familiar surroundings so, if you can, find someone dependable to take care of them in their own environment.

Leaving instructions

If your chickens need cleaning out or extra care while you are away, make sure your chicken sitter is willing and able to undertake this.

Be fair to your chicken sitter, so that both of you can relax. Always leave more than enough feed and bedding. Write out a list of what needs to be done – verbal instructions can be confusing and easily forgotten. Include details of the hens and any individual quirks ('white hen is rather nervous'). Add information about what to do if a bird is unwell, along with the phone number of your vet.

Key Points

- Make sure the chickens have food, water and grit every morning
- Check the birds as they come out in the morning, and ensure fencing is functional
- Shut chickens securely in their house at night and remove all food from the run
- Don't put feeders and drinkers in the henhouse overnight
- Collect eggs regularly
- Check drinkers in hot or freezing weather
- Keep the run clean – shift movable runs frequently
- Clean the henhouse weekly and scrub it occasionally
- Look for signs of red mite when cleaning the house
- Wash feeders and drinkers regularly
- Deal with repairs to the henhouse promptly and make sure it is sound before winter
- Check electric fencing frequently and trim surrounding vegetation
- Inspect your chickens weekly
- Internal worms can badly affect chickens' health and may prove fatal
- Flubenvet worming powder is mixed with feed for seven days
- If the vet prescribes an alternative wormer, there will be an egg-withdrawal period when you cannot eat the eggs produced
- Chickens should be wormed at least every six months – more often if necessary
- Repeat worming treatment when dealing with an infestation
- Herbal worm products are available, but don't kill gapeworms
- Maintain hygiene to help prevent a build-up of worm-eggs
- Change the run material and move electric fencing whenever necessary
- The flight feathers on one wing can be clipped to stop chickens escaping
- Re-clip after moulting, but make sure feathers are fully grown first
- Arrange for your chickens to be cared for when you are away from home

KIDS' CORNER
Quiz
Now you've read Chapter 8, test your chicken-keeping skills.

QUESTION ONE
How often would you normally clean out the henhouse?
(a) Once a week
(b) Once a month
(c) Once a year

QUESTION TWO
How would you get hold of a chicken that doesn't want to be caught?
(a) Run after it as fast as you can
(b) Wait until the chicken is roosting, then lift it off the perch
(c) Ask all your friends to help you round it up

QUESTION THREE
What is 'worming'?
(a) Digging up worms to feed to the chickens
(b) Giving the chickens medicine to kill any worms living inside them
(c) Removing worms from the chicken run to stop the chickens eating them

QUESTION FOUR
Which of these statements is wrong?
(a) Chickens should be regularly checked for injuries or illness
(b) The henhouse should be scrubbed with detergent occasionally
(c) You must always clip your chickens' wings

QUESTION FIVE
Before you go on holiday, should you:
(a) Arrange for someone to check the chickens every day?
(b) Buy some automatic equipment to look after the chickens while you are away?
(c) Leave the chickens with enough food and water to last them a few days?

How did you do? Make sure you know everything about caring for your new chickens!

Chicken Chat

'To clip someone's wings': In this chapter you read about clipping chickens' wings to stop them from flying. When we talk about clipping a person's wings, it means limiting their powers. For example: 'Our Head Girl is very bossy – it's time her wings were clipped!'

Chicken Joke

If a motorway is a fast road for cars, what's a henway?
About two kilos!

Something to Do . . .

Cleaning out your chickens is a very important part of keeping them healthy. A dirty henhouse can lead to diseases or breathing problems. A regular clean also gives you a chance to check the house for red mite – this parasite can make your chickens very weak or even kill them.

Cleaning the chicken house isn't everyone's favourite job, but put on your old clothes and rubber gloves, and do your hen-housework!

Why not work out a daily, weekly and monthly routine for looking after your chickens? Write it all down to make sure you don't forget anything!

CHAPTER NINE

Outfoxing the Fox: Predators

The fox is the best known of chicken predators, but unfortunately it is not the only one. Many other creatures like the taste of chicken, chicks or eggs.

The previous chapters have mentioned the importance of security – these are the creatures you are trying to keep out.

The Fox

Foxes have been drawn to towns and cities by the bright lights, and many are living it up on discarded takeaways, while being photographed and admired in urban gardens. Their country cousins dodge bullets and survive on beetles. It's not surprising that so many foxes are re-locating.

Town or country, if you have chickens there is a good chance that a fox will know about it. If you have never seen a fox around that doesn't mean there aren't any – just that you have not been worthy of their attentions before.

Traditionally foxes like to hunt at night, but they aren't all traditionalists, and are often around during the day – especially the city types. Country foxes may also vary their hours of work, so nobody can afford to be complacent.

Foxes are skilled and resourceful predators. They can scale surprisingly high fences, squeeze through tiny gaps and burrow like rabbits. If you want free-range chickens in an area where daytime foxes are operating, you will need to use electric poultry netting.

There are various electrical gadgets and sprays sold to deter foxes – male human urine is also said to be effective! The success of these measures depends largely on the determination of the individual fox. Although most animals are suspicious of anything strange at first, once they have become used to it they often take no more notice.

Foxes are naturally wary of people, but urban foxes have become more confident, and country foxes may also take chances. A cockerel was snatched from almost under my nose while I was cleaning out the henhouse – luckily I was able to 'persuade' the fox to let him go!

If you have a really persistent fox, there are pest-control companies who will shoot the offender – you will have to pay for this service. In rural areas, a farmer or gamekeeper may be able to help.

Don't underestimate foxes. Their main preoccupation is finding food, and they have got twenty-four hours a day to do it. You will need all your wits about you to stay one step ahead.

The Badger

Badgers also take chickens, and they are creatures of habit – once they have found food they will return every night.

Although less agile than foxes, badgers can climb quite well, and their digging ability is second to none. With sharp claws, mighty jaws and formidable strength, the badger can be a fearsome killer.

Fortunately, badgers are mainly nocturnal (although there have been occasional daytime sightings). They are protected, but if you have a problem you could try contacting your local badger conservation group for advice.

You will need to keep your chickens very secure at night if there is a badger sett nearby. Make sure the house is really sturdy – imagine it withstanding an attack by a large, hungry animal. Badgers have been known to rip off doors and even complete wood panels to get at chickens. Exterior nest-boxes and pop-hole doors are often weak points.

Mink and Otters

Mink were brought to Britain from North America to be bred on fur farms. Some escaped, some were released by protestors, and mink have been breeding in the wild since at least the 1950s.

Mink are widespread throughout Britain except for a few mountainous regions. They live by water, where they cause extensive damage to the native water voles. Mink are very territorial creatures, and two are rarely found in the same area except during the breeding season.

Reasonably large (males weigh about 1kg and are around 60cm long with a 15cm tail) and absolutely fearless, the mink has no natural predators. It will kill anything it can overpower, even if it is larger than itself.

Mink will take waterfowl as well as fish – they are excellent swimmers – and often leave the riverbank to look for rabbits. They climb well too, and will annihilate a chicken run if they gain access.

Watch out for mink if you have water nearby. They tend to explore rabbit warrens, so stop up any burrows in the vicinity.

Any mink caught in a trap must be humanely despatched. It is an offence to release mink back into the wild.

My Mink Attack

Living close to a river, I have probably been 'lucky' to have only experienced one mink attack. I was also fortunate enough to return home during the killing spree – in time to see and hear one of my hens being slaughtered. The mink completely ignored my yells, and it was only by beating it off with a lunge whip that I saved my buff Orpington cockerel. All that was left of a pretty little Pekin cockerel was a pile of grey feathers on the riverbank. We had a return visit when the mink took one of our youngsters, but once we had invested in a mink trap it left us alone. Telling the local fishing syndicate about its presence may also have helped.

The mink is related to the otter, which is a much larger animal with a flatter tail and broader face. It is making a comeback in Britain's rivers, and may also be found around some rocky coasts. Although they mainly eat fish, otters have been known to take waterfowl and poultry if the opportunity presents itself, and they are certainly strong enough to do some damage. Otters are generally shyer than mink and more likely to be nocturnal (apart from coastal otters). They are a protected species.

Stoats, Weasels, Polecats, Ferrets and Pine Martens

Stoats are larger than weasels, have a black tip to their tails, and some turn white in winter, when they are known as ermine. It's not easy to tell them apart, but both animals are a danger to chickens and can get through incredibly small openings. They can be caught in traps.

Wild polecats and escaped domestic ferrets are efficient killers, although polecats are fairly rare. The two species can interbreed. Ferrets are less likely to be wary of people, but usually don't survive long in the wild. Both these creatures can get through very small gaps.

Pine martens are mainly found in Scotland and Ireland, although there have been occasional sightings in upland areas of England and Wales too. About the size of a small cat, these attractive animals certainly pose a danger to chickens, although they are mainly nocturnal. They tend to enlarge holes in rotten wood, rather than make new ones, and are excellent climbers. Electric netting is a good defence, and an electrified overhang around the chicken run may be needed for a persistent offender. The pine marten is a protected species.

Birds of Prey

Some birds of prey will take chickens – especially small breeds or youngsters. The chickens usually raise the alarm if a hawk is around, and free-range birds will try to hide under bushes. A roofed run can protect enclosed chickens.

Crows, Magpies, Jackdaws and Rooks

These birds will steal chicks and eggs. Very small bantams could be at risk too. Wild birds also spread disease.

This section of the bird family is known for intelligence, so don't tempt them with easily accessible food or uncollected eggs. Pinning strips of plastic (cut-up black bin-bags, for example) along the tops of the pop-holes should help keep them out of the henhouse. The hens will push through the strips, but the darkened coop will be less inviting to wild birds, who will also find the movement of the plastic disconcerting, while the eggs will be less visible.

However, there is an advantage to having some of these birds around – they will give an early warning of hawks and will chase them off too.

Rats and Squirrels

Rats can attack sleeping or sitting hens, but are more likely to take eggs (at which they are very adept) and kill chicks. Foxes eat rats and may be attracted by their presence.

Rats often move in during the winter when food is scarce. Make food and water as inaccessible to them as possible, being sure to clear everything away at night.

A pest controller told me that if you see a rat in daytime, you have got a problem. The senior rats feed at night, but if there's not enough to go round, the younger ones have to venture out in the daylight.

If you see a rat at any time, or simply spot evidence of their presence (droppings, small holes or gnawing), it's essential to act before they breed more rats. You can trap or poison them – make sure children and pets can't access the poison, and also that your dog or cat doesn't eat poisoned rats. Contact the local pest control officer for advice.

Rats carry diseases that can seriously affect humans. If you touch any areas where rats may have been, thoroughly wash your hands (if the skin is broken or if you are bitten by a rat, seek medical advice immediately). Wear gloves if you have to deal with dead rats and wash your hands afterwards.

Grey squirrels also like chicken eggs, and may take chicks too. Full-grown hens will usually see off a squirrel, but young birds or small bantams could possibly be attacked.

Trapping Responsibly

It is illegal to inflict suffering on any animal, so traps must be checked at least twice a day, and any creature captured should be disposed of humanely. Releasing an animal elsewhere can result in welfare issues, and is illegal in the case of non-native species such as grey squirrels and mink.

Domestic Cats and Dogs

Although cats don't often pose a threat to full-size, adult chickens, they will take chicks, and small birds may also be at risk.

Chickens can bring out the hunting instinct in even docile, well-behaved dogs, so make introductions very cautiously. Some dogs happily take to chickens and may even protect them, but no dog should be trusted until you are completely sure. Dogs can often be trained to live in harmony with chickens, but this may take some time and effort.

Be careful of neighbouring dogs too, and make sure your flock is protected if you live near a footpath.

Humans

Chicken rustling is becoming ever more common, and isn't restricted to rare or expensive breeds. Equipment may also be taken – sometimes the whole coop with its occupants. While chickens on allotments or those kept in paddocks are more vulnerable, thieves have also struck in back gardens. A stout padlock and chain on the chicken-house is a good start, but your local police will be able to help with expert advice on security.

Keeping Your Chickens Safe

This may sound impossible after the foregoing list of predators, but many of these animals aren't actively out to grab your chickens. Most are opportunists who may stumble across the chickens while in pursuit of their normal diet. Not providing these creatures with a reason to visit is easier than trying to eradicate them once they have discovered an easy food source.

Key Points

Here are some tips for protecting your chickens:

- Make sure your henhouse and run provide adequate protection from predators
- No matter how secure your fencing, always shut your chickens in their house at night
- Don't leave food and water out overnight
- Regularly check your henhouse and run for weak spots, digging, rotten wood and signs of gnawing
- Animals are attracted by food and water sources – including compost heaps, ponds, fallen fruit and bags of rubbish
- Foxes can manage high fences – and can manage even higher ones if there is something for them to stand on
- Clear undergrowth around chicken housing to eliminate ground cover
- Discourage rabbits, and stop up rabbit holes around chicken housing
- Act quickly if there is any evidence of rats
- Bear in mind that some creatures can squeeze through minuscule gaps
- Security lights may help deter night-time predators
- Remember that foxes aren't the only predators of poultry
- Domestic pets can also be a threat to chickens
- Collect eggs frequently so they don't attract unwanted attention
- Make sure outbuildings are secure from two-legged thieves

Free-Range Chickens

Although free-range chickens may appear to offer an easy target, they quickly learn to be alert to danger. Leaving their wings unclipped gives them a better chance of evading predators.

If a predator gets into a chicken run, all or most of the hens are likely to be killed. Free-range chickens may only suffer occasional losses (provided they are securely shut in at night), but if this turns into a steady stream, you will have to think again.

KIDS' CORNER

Quiz

What did you learn about chicken predators from Chapter 9?

QUESTION ONE
Which of these statements is true?
- (a) Foxes only live in the country
- (b) Foxes are only around in the night-time
- (c) Foxes aren't the only threat to chickens

QUESTION TWO
Which of these animals isn't a predator of chickens?
- (a) Deer
- (b) Badger
- (c) Stoat

QUESTION THREE
Which of these chicken predators is a protected species and must not be harmed?
- (a) Weasel
- (b) Mink
- (c) Pine Marten

QUESTION FOUR
Why might you need to pin strips of plastic over the pop-hole?
- (a) To keep out rats
- (b) To stop birds coming in to steal the chickens' eggs
- (c) To frighten off foxes

QUESTION FIVE
What should you do if you find you have rats?
- (a) Take no notice – they will do no harm
- (b) Don't tell anyone in case you are stopped from keeping chickens
- (c) Wash your hands thoroughly and tell an adult

ANSWERS
One (c); Two (a); Three (c); Four (b); Five (c)

Knowing about predators will help you to protect your chickens and their eggs.

Chicken Chat

'Like leaving a fox to guard the henhouse': It wouldn't be a good idea to leave a fox to look after your chickens! This saying is used when someone is given a task that they are likely to turn to their own advantage. For example: 'Katie can't resist chocolate – letting her run the sweet shop is like leaving a fox to guard the henhouse'.

Chicken Joke

Which dance do chickens hate?
The foxtrot!

Something to Do . . .

Make it your job to regularly check the chicken house and run for any weak spots. Imagine you are a big hungry animal – could you get in? Now pretend you could make yourself thin enough to slither through a little gap – can you see any spaces where you might slip through? Look out for any signs of digging and gnawing too.

If you can't fix the problem yourself, ask an adult to help.

CHAPTER TEN

You Can't Make an Omelette Without Breaking Eggs: Eggs and What to Do with Them

The new chickens are happily settled and you're pacing the floor, waiting for that first egg. It may take a while to arrive, but when it does it can seem like a small miracle. In some ways it is ...

Creating the Wonderful Egg

Female chicks hatch with a lifetime of potential eggs inside them. When a pullet reaches maturity, a yolk is released into the reproductive system. Egg white (albumen) is added, followed by the shell membrane and the shell. Finally the egg gets a protective coating (the cuticle).

This all takes around twenty-five hours, of which twenty are taken up with shell production. When a hen is in full lay, the process starts again within half an hour.

An egg a day?

Good layers can manage a daily egg for several days, but eventually there will be a gap. As it takes twenty-five hours for an egg to develop, laying will occur a little later each day. Egg production is stimulated by daylight, so there will come a point when the next yolk is not released until the following morning.

Around fourteen hours of daylight are required for maximum egg production – most hens stop or slow down in the short days of winter.

Artificial lighting can be used to extend the day, but is unlikely to be cost-effective for a small flock, and the continuous, intensive production of eggs may lead to laying problems (see Chapter 11).

If artificial lighting is required, it's easier to have it come on early in the mornings rather than lengthening the afternoons. Evening lighting requires a dimmer switch to imitate twilight, so that the hens go to roost naturally and aren't left on the floor in the dark.

Eggs forever?

Bearing in mind that a hen's egg supplies will eventually run out, many poultry keepers let their birds take a natural break over the winter. Rather than putting all their eggs in one basket, they extend the number of seasons in which their hens are likely to lay.

The first year of laying is usually the most productive. There will be fewer eggs in the second and subsequent years, but they are often larger. Eventually the hen will lay only an occasional egg or will stop altogether.

Does size matter?

A pullet's first eggs are usually small as her system adapts to regular laying. Eggs should become larger as time goes on.

The biggest hens don't inevitably lay the biggest eggs. Some giant breeds supply rather small eggs, while there are slender light breeds (such as the Leghorn) that lay surprisingly large eggs. Although bantams are much smaller than their standard counterparts, this isn't necessarily reflected in the size of eggs they produce.

Bantam eggs can be particularly rich and tasty, but you will usually require more of them when following a recipe. Smaller eggs can easily be used for frying (just have two!), scrambling or omelettes, but large eggs tend to be preferred for boiling and are generally more popular.

Coats of many colours

Eggshells vary from the darkest brown through to speckled, light brown, tinted, cream, white, blue or green – with several different hens you could produce a multi-coloured egg-box. There is no difference in taste or nutritional value, but many people prefer eggs of a certain colour.

James Bond's Favourite Egg

James Bond enjoyed the dark-brown Marans' eggs for breakfast at home in London. They were supplied by a friend of May, his Scottish housekeeper.

Colour is usually added at the end of the egg's development – it can actually be rubbed off a newly laid brown egg. Shells are white inside, apart from the blue/green eggs where the colour goes right through.

Shell colour can vary depending on the hen's breeding strain, and often fades as the laying season progresses, especially if the hen is a good layer. Other reasons for loss of shell colour may be poor nutrition, stress, hot weather, age, disease or parasites.

A hen's ear lobes are often believed to indicate shell colour – white suggesting white eggs, and brown eggs from hens with red lobes. This is sometimes the case, but there are several exceptions.

The eggshell

Whatever colour the shell, its purpose is to protect a developing chick. It is mainly composed of calcium carbonate and is surprisingly strong.

The shell is porous to allow it to 'breathe' and to lose moisture as the chick grows. On a newly laid egg the cuticle covering the shell is often damp, but quickly dries to a protective sheen. Washing removes the cuticle and can allow bacteria to enter.

Soiled eggs can sometimes be cleaned by brushing gently with an old toothbrush. If washing is essential, use water slightly warmer than the egg. This will make the shell membrane expand and block the pores. Cold water will shrink the membrane and draw in bacteria. Eggs that have required cleaning should not be offered for sale.

It's better to keep eggs from becoming soiled in the first place. Make sure nest-boxes are clean, and prevent hens from roosting in them. Try to keep the area surrounding the henhouse dry so the hens don't bring muddy feet into the nest-boxes. Collect eggs regularly, making sure the container you put them in and your hands are clean.

Inside the shell

The eggshell is lined with an outer and inner membrane, incorporating an air space, which is usually located at the wider end of the egg. As the egg ages, moisture is lost from the egg and replaced by air – the warmer and drier the atmosphere, the faster this happens. The height of the air space indicates the age and quality of an egg.

In a fresh egg you can easily see two kinds of albumen (white) – a thin outer layer and a thick jelly-like layer around the yolk. As the egg ages, all the white becomes watery. If a fresh egg has a watery white it may be due to hot weather, the age of the hen or exposure to disease, such as infectious bronchitis.

If the egg is very fresh you may see the chalazae – white string-like substances that hold the yolk in place. These disappear with storage and don't affect the taste of the egg.

In the middle of the egg sits the prize – a golden yellow yolk. The colour comes from greens (especially grass) in the hen's diet, although there are feeds which include artificial colour enhancers. Chickens free-ranging on grass in summer will produce yolks of a colour and flavour undreamed of by battery hens. Greenish yolks can be caused by the hen eating acorns or weeds like shepherd's purse.

On the edge of the yolk is the germinal disc – a small white spot containing a single female cell. If the egg is fertile, this spot will be larger and rounder. Fertile eggs are fine to eat as long as incubation hasn't started, in which case red veins may be seen when the egg is broken. This can happen within a couple of days if the eggs have been sat on by a broody hen or are left in very warm conditions.

Finding Alien Eggs

Funny things can happen during the egg's long journey to the outside world. Some are just glitches in the system, although odd eggs may occasionally indicate a problem.

Tiny eggs

Sometimes a mini egg appears, usually without a yolk.

This is often caused by a piece of tissue that has separated from the ovary and been mistaken for a yolk and encased with egg-white and shell. It is quite common with pullets or older hens at the end of their laying season, and is usually nothing to worry about.

These little eggs have many names: witch eggs, fairy eggs, wind eggs, cock eggs – the last because it was once believed that they were laid by cockerels!

The Monster from the Egg

In ancient legend tiny yolkless eggs were laid by cockerels. Should one be hatched by a toad, it would grow into a dragon-like creature that could kill people with a single glance. The way to avoid such a fate was to throw the egg over the house, being careful it didn't touch the roof.

Eggs with multiple yolks

Sometimes more than one yolk is released during production – either together or in quick succession – and they become enclosed in one shell. Multiple yolks are more likely in pullets, although this can also be an inherited feature.

Double and triple yolks are fairly rare – nine yolks in one shell is claimed to be the record.

These eggs are likely to be larger, and there may be blood on the shells. If a hen frequently lays huge eggs, it's worth checking that she hasn't suffered any internal damage.

It is very rare for eggs with multiple yolks to hatch, as there isn't room for two chicks to grow in the same shell.

Soft, thin or wrinkled shells

Sometimes you pick up what looks like a normal egg, only to find a jelly-like substance without a shell. I hate this! Or the shell may be wrinkled or very thin. These shells, or lack of them, can encourage egg-eating, so they should be taken away as quickly as possible.

An occasional shell-less or wrinkled egg may be the result of a sudden fright. Shell problems can also be due to age – pullets starting to lay or older hens at the end of their laying years. Shell quality also sometimes deteriorates as hens start moulting.

If there isn't an obvious cause, and the situation doesn't resolve itself, then further investigation might be needed. Dietary imbalance, stress or disease can affect shell quality.

Dietary imbalance

It shouldn't be necessary to give extra calcium if the hens are eating a good layers' feed and have access to soluble grit – an excess of calcium can cause health problems. A better plan is cutting out treats to make sure the hens are eating properly. Check they are getting enough vitamin D, which enables the utilization of calcium. Feeds should contain vitamin D, but it is also obtained from sunlight. Cod liver oil can be given as a supplement to hens kept mainly under cover.

Hot weather can result in hens eating less food. Make sure they have access to shade and cool water.

Occasionally a hen is unable to process calcium properly, but veterinary advice will be required to establish if this is the case.

Stress

Stress can be caused in a number of ways: a change in routine, bullying by other hens, overheating etc. (see Chapter 11).

Parasites and disease

Shell quality may be affected by parasites or disease, but there will usually be other symptoms.

Exposure to infectious bronchitis can cause damage to the reproductive system, and the hen will only lay soft-shelled or wrinkled eggs. There is nothing that can be done about this, and the hen should be culled as she is likely to be a carrier of the disease.

Misshapen eggs

Sometimes eggs have an extra band of shell around the middle or a dusting of powdery calcium. There have even been cases where a whole egg is encased in another shell!

Often the result of stress or shock, these eggs are more likely to be produced by pullets or old hens. There is usually nothing to worry about, unless the condition persists. Try to keep the flock as free from stress as possible.

Blood on the shell

There may be a smear of blood on an eggshell, especially when a young hen is starting to lay, or if the egg is particularly large. Keep an eye on the hen, but unless there is a lot of blood, or blood around the vent, there won't usually be a problem.

Small red dots on the shells can indicate red mite.

Meat spots and blood spots

These are seldom seen in shop-bought eggs, as they are routinely scanned before packing.

Blood spots may occur at the edge of the yolk, while a small brown speck in the white is known as a meat spot. They are produced when the yolk breaks away from the ovary and are harmless, although they don't look very appetizing.

Meat or blood spots may be a result of age and are often caused by shock. Some hens regularly produce eggs with these imperfections, which may be an inherited trait, so it would be best not to breed from them.

Where Are the Eggs?

Maybe you're not getting any eggs at all. Could something be wrong?

It might be, but then again all could be well. There are several reasons for a lack of eggs:

- Time of year: many pure-breeds stop laying over the winter and the pullets may not start until the spring, depending on how early in the year they were hatched.
- Age of hen: pullets mature at different ages depending on their breed. Old hens will eventually stop laying – hybrids often stop earlier than pure-breeds.
- Moulting: hens don't lay eggs during the moult and re-feathering (see Chapter 11).
- Diet and water: incorrect feeding (unbalanced diet, excess of treats) and obesity can result in fewer eggs. Lack of water may cause egg production to cease for several days.
- Broodiness: broody hens don't lay eggs.
- Parasites: an infestation of any kind will debilitate the hens – laying will slow down or stop.
- Disease: hens will stop laying if they are ill but there will usually be other symptoms too.
- Laying outside: free-range hens may make nests under bushes, in clumps of nettles or some other unsuitable spot.
- Egg eating: sometimes hens eat their own eggs (see next section) but you will probably notice shell fragments.
- Rats and other predators: rats will take eggs and can cleverly roll them away. Birds such as crows and magpies will steal eggs. Grey squirrels, weasels and stoats enjoy eggs and may take chickens too (see Chapter 9).

Egg eating

Chickens don't realize how delicious their eggs are – and it's best to keep it this way. Egg eating is a difficult habit to break. Collecting eggs regularly removes temptation. If thin shells are leading to broken eggs in the nest-box, try to identify the reason and improve matters.

Nutritional deficiencies and lack of water can encourage hens to investigate their eggs, especially in the summer months.

If you have an egg-eater, putting pot eggs (imitation eggs) or golf balls in the nest-boxes may make her decide that eggs aren't always tasty. Replacing the contents of an egg with something unpleasant tasting has variable results, but adding some vegetable dye may help identify the culprit.

Nest-boxes should be in the darkest part of the house – make sure there are enough for all the laying hens. Pinning black plastic strips along the top of the nest-boxes keeps eggs away from greedy eyes. There are also 'rollaway' nest-boxes, where the eggs roll into a collection tray out of reach of the hens.

Knowing if a hen is ready to lay

A firm red comb is often the most obvious sign. The hen may also start 'talking' – making soft clucking sounds.

Not all breeds have visible combs, so a more accurate method is to check the width of the pelvic bones. These can be felt either side of the vent, and will be close together if the hen isn't laying. When she is ready to start they will be about two fingers apart, and will widen to around three fingers (5cm) as laying increases.

Eggs-Travaganza! Dealing With All the Eggs

Your patience has been rewarded – that first egg could hardly be more precious if it was made of solid gold. You may be wondering whether to eat it or frame it!

There's a kind of wonder in collecting freshly laid eggs every day, but when your hens get into their stride the wonder may turn to panic as the fridge fills with egg-boxes at an alarming rate.

When I was a child we had eggs for breakfast every morning, but cooked breakfasts are much rarer these days. Although home-baking is making a comeback, four hens in full lay can easily provide over twenty eggs a week, which is an awful lot of cakes.

The spring abundance will diminish as the year wears on and, come winter, eggs will be precious again. The question is, how to make the most of them during the months of plenty, while preparing for leaner times to come.

Storing eggs

There is some debate about whether eggs should be stored in the fridge, but they will deteriorate more quickly in the warmth of a kitchen. Unless you are lucky enough to have an old-fashioned cool pantry or cellar, refrigeration is the safest option for keeping eggs fresh.

Egg-boxes are the perfect containers and can be bought from chicken supply stores. Place the eggs pointed-end-down so the air-space can 'breathe'. Number or date the boxes so you know the age of the eggs inside – this will be helpful if you have a springtime glut.

Eggs can absorb smells, which may be fine if you have a basket of truffles, but not so great if they are stored next to onions or smelly cheese.

Ideally eggs should be taken out of the fridge half an hour before you use them.

You'll have noticed that eggs aren't refrigerated in the shops, although the boxes advise customers that eggs should be chilled after purchase. This is a legal requirement to prevent changes in temperature during transportation.

Preserving eggs

No shop-bought eggs can compete with those laid by your own hens, so it makes sense to preserve as many as possible when you have an excess.

Freezing

Eggs can't be frozen in the shell (they would crack), but they can be stored in the freezer in a variety of other ways. Choose only fresh, undamaged eggs.

The easiest method is to freeze eggs in quantities you regularly use (e.g. three for an omelette). Break the eggs into a freezer-suitable container and mix very gently, without making them frothy. Add a pinch of salt or a little caster sugar, depending on their intended use – don't forget to label the container! Eggs will keep up to six months like this.

You can also freeze yolks and whites separately, which is handy if you are making meringues and want to keep the yolks for another day. Whites can be frozen as they are. Yolks should be lightly whisked with a little salt or caster sugar to stop them from thickening and gelling.

The whisked yolks can be frozen in ice-cube trays, then popped out and stored in a freezer bag. One cube will equal a standard yolk. Whites can also be frozen like this – two cubes will equal a standard white.

If you enjoy baking (and chicken-keeping can turn people into bakers), consider freezing cooked dishes such as quiches and cakes. Sponge cakes made with eggs from your own hens are especially delicious. They are a useful standby in the freezer too, enabling you to put a pudding or homemade cake together with very little effort.

Boiled eggs don't freeze well – the whites go rubbery and the yolks crumble.

Pickling

Pickled eggs are very easy to make. They keep for ages without refrigeration, and make a tasty addition to salads and snacks. You will need a large preserving jar. The 'Kilner' type with glass lids are ideal – metal lids will react with the vinegar. First sterilize the jar and lid – either put them through a very hot cycle in the dishwasher, or wash and rinse thoroughly, then heat for 10 minutes in the oven at 170°C /fan 150°C /325°F /Gas 3. Allow to cool before using.

For a dozen eggs, simmer a litre of vinegar with a few whole spices (or a bag of pickling spice) for around ten minutes. Vinegar and spices can be varied to taste. Cider and wine vinegars are milder, or you can dilute malt vinegar with half as much water. A tablespoon of sugar makes a sweeter pickle, a chilli will add some bite, or you can add a few cloves of garlic. Experiment with small batches of eggs until you find the magic formula.

Choose eggs that are more than a week old – they will be easier to shell. Hard-boil them thoroughly, plunge into cold water, then shell them and place in the jar. When the vinegar has cooled, strain it over the eggs so they are completely covered. Seal and leave for about two weeks.

Preserving Eggs in Waterglass

In the days before refrigeration, eggs were preserved in a solution of sodium silicate or waterglass. This seals the shell pores, keeping bacteria out and moisture inside.

Waterglass (not 'isinglass', which is used for wine-making) can be bought online, and the chemist may also be able to order it. One part waterglass is mixed with nine parts boiled, cooled water. This is poured into a lidded container – plastic, enamel or earthenware – which has been thoroughly cleaned and scalded. Remember to leave enough room for you to add the eggs.

Use only clean, unwashed, fresh eggs, and carefully check for any cracks. Infertile eggs are said to keep better than fertile ones. Place the eggs pointed-end-down, and make sure the solution covers them by about 5cm. Keep the lid on the container to help avoid evaporation and top-up when necessary; store in a cool, dark place.

Wash the eggs in cold water before using, and prick with a pin before boiling. An undetected hairline crack will cause an egg to go bad, so each egg should be broken into a cup before being added to a dish.

My father remembers eggs being preserved this way in the 1930s. He says they emerged as good as fresh, although they would probably be unsuitable for certain dishes. As the eggs age, the yolks become more delicate, and the whites become thinner and difficult to whisk.

Giving eggs away

Maybe this sounds wasteful – and hardly making the most of all your hens' efforts. It depends who you give them to. A box or two may help keep the neighbours happy, especially if they have been disturbed, or might be disturbed, by your chickens.

Remember those who have helped too. My hen-sitting friend is convinced that our hens keep a calendar so they know when we are going away – not a sniff of an egg in recompense for her twice-daily visits. When we have an excess, I make sure she shares in the bounty.

A box of fresh eggs is usually an acceptable gift – maybe instead of chocolates or wine when visiting friends. You can amaze the recipients by telling them the name of the hen that laid each egg!

Selling eggs

When you have paid off all your obligations and the family is refusing to eat another omelette, you can sell surplus eggs. As eggs are a food product, there are some regulations, but these are fairly straightforward for small-scale flocks.

Legislation may vary in different parts of the British Isles and can be subject to change. For current advice you should check with DEFRA or the Animal and Plant Health Agency (Egg Production and Marketing) – see 'Further Reference'.

At present the rules are:

- You must register with DEFRA if you have more than fifty birds, irrespective of whether you intend to sell eggs.
- There is no need to register with the APHA as an egg producer if you have less than 350 birds, and only sell eggs direct to customers for their own use. This may be from the farm (or garden) where they were laid or via door-to-door sales.
- Unregistered producers must not grade their eggs for size or quality (large, medium, small, Class A or B).
- Eggs may only be described as 'free-range' or 'organic' if they meet the relevant criteria.

Selling eggs at a local market

If eggs are sold by the producer at a local market, further regulations apply.

- If you have more than fifty birds, you must register as an egg producer with the APHA.
- If you are not registered, eggs must not be size- or quality-graded.
- Your egg-boxes must display the following information: name and address; best before date (maximum twenty-eight days from laying); advice to keep eggs chilled after purchase.
- Some markets also require eggs to be stamped with a producer number – in which case you will have to register with the APHA.

Larger-scale selling

If you wish to sell eggs to shops, restaurants or hotels, you will need to register with the APHA regardless of the number of hens you keep, and have the eggs graded at an egg-packing centre. It's also possible to set up a small packing centre at home – details and forms are available on the APHA website. There's no charge for registering as an egg-producer or packer, but you'll need to provide some equipment for packing. This includes 'candling' apparatus, which allows you to inspect the egg's contents by means of a bright light.

Eggs for sale!

In the early stages of chicken keeping you may only be selling a handful of surplus eggs, rather than registering as an egg producer. However, you will still want to supply your customers with eggs that are as good as, if not better than, those produced commercially – if for no other reason than to encourage them to return for more.

While you mustn't grade your eggs if you aren't registered, it may be interesting to know some of the quality standards applied to Class A eggs (i.e. those that are sold in the shell to consumers):

- The shell and cuticle must be of normal shape, clean and undamaged
- The air space height cannot exceed 6mm (4mm if marketed as 'extra fresh')
- When candled, the yolk should appear as a shadow with no clear outline
- The yolk should be slightly mobile when the egg is turned, returning to a central position
- The white must be clear and translucent
- The germinal disc must not be developed
- Foreign matter or smell is not permitted
- Eggs must not be washed or cleaned
- Eggs may not be treated for preservation or chilled in artificial temperatures of less than 5°C
- Eggs must always be kept clean, dry and out of direct sunlight
- Class A eggs are to be graded and sold according to weight

Eggs that don't meet these criteria may be disposed of in accordance with regulations or sold to industry for processing.

While you may not be able to check inside the eggs for defects, following a few simple guidelines will help maintain good quality:

- Only sell eggs that have come from healthy hens, kept in clean conditions, and fed a good diet.
- Remember that shocks and upsets to a laying flock can cause egg defects for a few days afterwards.
- Eggs should be clean but should not have been brushed or washed.
- Sell only eggs with perfect shells – no odd shapes, lumps or cracks.
- Eggs should be fresh. Collect them at least daily, store in a cool place and sell within a day or two of laying.
- Don't sell eggs from hens that regularly produce watery whites, meat spots or blood spots.
- Many people don't like eating fertile eggs. It's also impossible to guarantee that the eggs will be stored correctly once purchased. If you have a cockerel with your hens, you should advise customers.

Making Use of Your Eggs

New chicken keepers usually find themselves eating more eggs than before. From being a useful standby in the fridge, eggs suddenly become a gourmet treat when you can have them on the plate within minutes of being laid!

A fresh egg will sink to the bottom of a glass of cold water

How fresh are the eggs?

Even if you check the nest-boxes regularly, you may discover a pile of eggs hidden under a bush. They can be tested for freshness – but even so should be hard-boiled or baked and not offered for sale.

Submerge the eggs in cold water. Fresh eggs will lay flat on the bottom of the bowl. Staler eggs will tilt slightly at the broad end, becoming more vertical as they age. An egg that floats is bad. Be careful about relying absolutely on this

An egg that floats is bad

test – eggs with hairline cracks can perform differently and incubation may have begun in fertile eggs.

You can also assess freshness by breaking the egg on to a plate. It should have a firm, plump yolk sitting up on a pool of thick, gelatinous albumen, with the thinner white surrounding it. The chalazae will be visible in a very fresh egg, and the albumen may be cloudy if the egg has just been laid. An older egg will have only thin, spreading white and a flat yolk.

How many eggs can you eat?

Like many foods, eggs have been in and out of fashion. The advertising slogan 'Go to Work on an Egg' suggested that an eggy breakfast was the ideal start to the day. Then there was the salmonella scare, as well as advice to eat no more than two eggs a week due to their cholesterol content. Research has now shown that foods containing cholesterol don't affect blood cholesterol levels as much as previously believed, and that saturated fats are the main culprits.

Moderation in all things is probably the best principle, but eggs pack protein, vitamins and minerals into their shells, and a medium egg is less than seventy calories.

Cooking with eggs

Most people have their favourite methods for preparing basic egg dishes, but the following hints and ideas may be useful.

Boiling

Eggs straight from the fridge can crack if plunged into simmering water – make a tiny hole at the rounded end to prevent this (egg-piercing gadgets make the operation easier, but you could also use a clean pin).

Hard-boiled eggs are easier to peel if they are more than a week old. Boil eggs for pickling thoroughly, but a slightly yielding yolk, still warm, makes a lovely sandwich. Plunging the eggs into cold water avoids a grey ring around the yolk. Unshelled hard-boiled eggs will keep in the fridge for a week or so.

Halved hard-boiled eggs can be covered with cheese sauce and browned under the grill for an easy supper dish (a layer of puréed spinach under the eggs is a good addition).

Poaching

For poaching in a pan of simmering water, the eggs must be really fresh or the whites will disintegrate into strings.

Scrambling

Whisk eggs well, melt a lump of butter in a small pan and cook slowly, stirring all the time. This is a good way of using bantam eggs.

Frying

Don't restrict fried eggs to accompanying bacon. Butter some good bread and enjoy a messy but delicious sandwich! Two small eggs are easier to cope with than one large one.

Omelettes

For me, one of the best omelette fillings is sorrel, which can easily be grown in the garden (as long as the chickens don't find it). Shred a few leaves and melt them in butter with a sprinkle of salt, adding them to the omelette when it is nearly set. Sorrel is rather like spinach, with a sharp, lemony flavour that complements eggs perfectly.

Pancakes

Try these little Scotch pancakes for a change. They can be eaten with a variety of accompaniments – butter, sugar, lemon, cheese, bacon, syrup, etc. Use 100g of self-raising flour, a pinch of salt, one egg and 150ml of milk to make a batter that isn't too liquid. Add spoonfuls to a medium hot pan that you've greased with very little butter or oil. When bubbles appear on the surface, turn the pancakes over and cook for another minute or two.

Pancakes can be made in advance and frozen.

Spaghetti alla Carbonara

This is a favourite dish of mine – easy, quick and delicious.

For two people use a 100g pack of pancetta, cut into thin strips and fried in butter. Beat two eggs with a little salt and pepper (add a spoonful of cream or an extra egg yolk to make it richer). Boil spaghetti or tagliatelle, drain it (not too thoroughly), return it to the saucepan and stir in the pancetta. Mix in the eggs, letting them just start to thicken. Serve immediately with grated Parmesan.

Mayonnaise

Picture yourself at the garden table with chickens clucking around your feet. Friends chat and sip wine, while you create homemade mayonnaise before their astonished eyes. Because it's made with fresh eggs, the mayonnaise will be golden rather than white – but it shouldn't be given to young children, the elderly or anyone else who is advised not to eat raw eggs.

Stir two egg yolks in a bowl with a pinch of salt (standing the bowl on a damp cloth stops it sliding around). Measure 150ml of light olive oil into a jug. Stir just one drop of oil into the yolks. Then stir in another drop. Carry on like this, and don't hurry the process or it will separate (if this happens, start again with a fresh yolk, adding the curdled mixture to it one drop at a time).

When about half the oil has been added, the mixture will have thickened. Add a few drops of lemon juice or wine vinegar. The remaining oil can now be added in a very slow trickle – keep stirring. Finally, taste for seasoning and add a little more lemon juice or vinegar.

Note: you can use other oils if preferred (extra-virgin olive oil gives a very strong flavour). A little mustard or a couple of cloves of garlic crushed into a paste can be added to the egg yolks with the salt.

Meringues

If you've made mayonnaise, there will be egg-whites to spare – or if you make meringues, you can use the yolks for mayonnaise. A sauce and dessert from two eggs!

Make sure the whisk and bowl are clean and grease-free. Use a glass or earthenware bowl, not plastic.

Separate the eggs carefully – the whites won't whisk properly if there's the slightest trace of yolk.

Heat the oven to 150°C (300°F/Gas 2). Line a baking sheet with non-stick baking paper.

For two egg whites measure out 110g caster sugar. Whisk the whites until they are stiff enough not to slide about the bowl when it's tipped – don't over-whisk, or they will collapse. Then whisk in the caster sugar, one spoonful at a time, until the mixture is stiff and shiny.

Fold in a few drops of lemon juice – this helps stabilize the mixture, but isn't essential. Pipe shapes or put spoonfuls of the mixture on to the baking sheet.

Put the meringues in the oven, reduce the heat to 140°C (275°F/Gas 1) and bake for one hour. Then turn off the heat and leave for several hours, until the oven is completely cold.

Peel off the baking paper and store the meringues in an airtight container. Meringues can also be frozen.

Custard and ice-cream

Proper custard is made with egg-yolks and milk or cream. Heat 300ml of cream until nearly boiling. Beat three yolks with a tablespoon of caster sugar and stir in the hot cream. Rinse out the saucepan, add the mixture and heat very slowly until it thickens, stirring all the time. This takes a while – cook it too fast and you'll have scrambled eggs! A teaspoon of cornflour beaten into the egg-yolks helps prevent this and speeds up the process too.

Custard is usually flavoured with vanilla – use vanilla sugar, infuse the milk with a vanilla pod or add a couple of drops of vanilla essence to the egg-yolks.

This custard is also the basis for the richest and creamiest ice-creams. Whipped cream is added when the mixture has cooled, and other ingredients can be incorporated to give different flavours and textures. This is a dinner-party luxury, rather than something to dish out in cornets on a hot day.

Don't forget the eggshells!

Eggshells have many uses:

In the garden

Crushed shells can be added to your compost heap or sprinkled around plants to protect them from slugs – I haven't found this particularly successful, but perhaps we just have tough slugs.

Shells can also be used to grow seedlings. Make a few holes in the bottom of a shell, fill it with potting compost and press in a seed. Stand the shells in egg-boxes to keep them stable. Plant the seedlings in their eggshells – crush the shells slightly first. This saves disturbing the delicate roots and the shells provide nutrients to the growing plants.

You can also grow cress in eggshells. Thoroughly wet a ball of cotton wool, put it in the shell and add some seeds. Stand the shell in an eggcup.

Even cooled water from boiling eggs is useful – use it for watering plants.

In the house

Add crushed eggshells to soapy water when cleaning awkward items like tall vases – leave to soak overnight to remove stains. Eggshells can also be used as a natural abrasive for scouring pans or kept in the sink strainer to trap solids – they will then clean the pipes as they disintegrate down the drain.

Add a crushed eggshell to your coffee-maker to reduce bitterness in coffee.

For artwork

Blow out the contents of the egg and decorate the shell (wash it first). Some poultry shows have classes for decorated eggs. Crushed shells can be made into mosaics.

For healing

The shell membrane has anti-microbial properties that are believed to help clear up cuts, spots, blisters and burns, as well as drawing out splinters. Peel the membrane from the shell (not easy) and apply it wet-side-down.

Key Points

- It takes about twenty-five hours for an egg to be produced
- Around fourteen hours of daylight are required for maximum egg production
- Use water slightly warmer than the egg if washing is essential
- As the egg ages, moisture is lost and replaced by air
- Tiny eggs without yolks are fairly common and usually nothing to worry about
- Shock or stress can cause misshapen eggshells
- Persistently wrinkled, thin or missing shells may indicate a problem
- A little blood on a shell isn't usually cause for concern; small red spots are a sign of red mite
- Meat or blood spots in the egg don't affect the taste
- Lack of eggs may be due to several causes – not all of them mean that something is wrong
- A firm red comb shows a hen is ready to lay – you can also check her pelvic bones
- Store eggs in the fridge or a cool place
- Raw eggs can be frozen but not in the shell
- There is no need to register with DEFRA if you only have a few hens and sell eggs from the gate
- If you are an unregistered producer, eggs must not be graded
- There are extra rules if eggs are sold at a local market
- Only graded eggs from registered producers can be sold to shops, restaurants or hotels
- Sell only fresh eggs with perfect shells that have not needed cleaning
- Research has shown that eating eggs doesn't affect blood cholesterol as much as previously thought

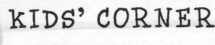

KIDS' CORNER

Quiz

Have your new hens laid their first eggs yet? While you're waiting, try this little eggsam!

QUESTION ONE

How long does it take for a hen to produce an egg?
- (a) Ten hours
- (b) Twenty-five hours
- (c) Fifty hours

QUESTION TWO

If you have to wash an egg, what should you use?
- (a) Cold water
- (b) Very hot water
- (c) Warm water

QUESTION THREE

What happens as an egg gets older?
- (a) Moisture is lost and replaced by air
- (b) Air is lost and replaced by moisture
- (c) The shell turns a different colour

QUESTION FOUR

If a hen has a firm red comb, what does this mean?
- (a) She is really a cockerel
- (b) She has stopped laying eggs
- (c) She is either laying eggs or will start soon

QUESTION FIVE

What are the rules about selling eggs to your neighbours?
- (a) If you only have a few hens, you can sell eggs direct to customers
- (b) You can only sell eggs to a shop
- (c) You will have to apply for a licence first

ANSWERS

One (b); Two (c); Three (a); Four (c); Five (a)

If you got all these right you are nearly an eggspert!

Chicken Chat

'Don't put all your eggs in one basket': If you put all your eggs in one basket and then dropped it, you would probably have nothing left but a nasty mess! This saying suggests that it's safer to spread your valuables around so that if you lose one you don't lose the lot.

Chicken Jokes

Did you hear about the hen that kept turning somersaults?
She laid scrambled eggs!

What is yellow and goes at 100 miles an hour?
A train-driver's egg sandwich!

Something to Do . . .

Break a fresh egg on to a plate and study it carefully. Can you see the thin and thick layers of egg-white? You may be able to see the chalazae (white strings that hold the yolk in place) as well. Look for the white spot on the edge of the yolk – the germinal disc.

Are there any blood spots on the yolk or meat spots on the white? Draw a diagram of the egg and label all the parts.

You can keep a chart of your hens' egg laying throughout the year, like the one on the next page:

Date	Jan	Feb	Mar	April	May	June	July	Aug	Sept	Oct	Nov	Dec
1												
2												
3												
4												
5												
6												
7												
8												
9												
10												
11												
12												
13												
14												
15												
16												
17												
18												
19												
20												
21												
22												
23												
24												
25												
26												
27												
28												
29												
30												
31												
Total												
Hens												
Eggs per Hen												

Notes

Each day write down how many eggs were laid, and then add them up at the end of the month. You can also work out how many eggs you got for each hen by dividing the number of eggs by the number of hens.

If anything unusual happened (broken eggs in the nest-box or a broody hen) make a note on your chart.

Your computer will do the sums if you enter this on a spreadsheet. You can even produce a graph to show how well your hens are laying.

This graph shows the eggs laid by a flock of hens over a period of four years. Can you see which months they laid the most eggs? Look at the year 2009 – the hens were laying less well as they got older.

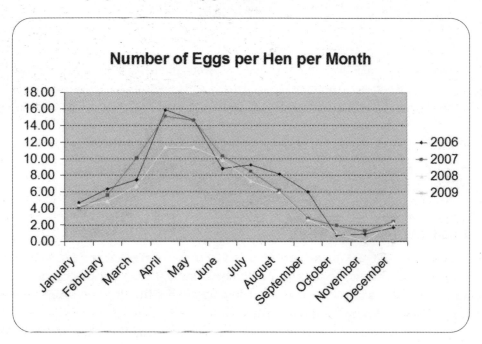

Number of Eggs per Hen per Month

In Fine Feather:
Dealing with Parasites and Ailments

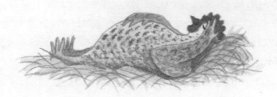

There is a wide variety of illnesses that can affect chickens, but many are never seen by the small-scale poultry keeper. Disease often results from overcrowding or poor living conditions, which can easily be avoided.

This chapter gives an overview of the various parasites, along with a few chicken ailments that may affect your flock. It is not intended to replace the advice of a vet; before bringing your first chickens home you should register with a local vet who has experience with poultry.

Looking for Problems

Regular checks (see Chapter 8) will help you to spot potential problems early. When a hen reaches the stage where illness is immediately obvious (hunched and miserable), she is likely to be in a bad way and difficult to save. She may have spread infection to other chickens too.

A sick hen may be bullied and even turned out of the flock to protect the health of the others, so chickens are good at hiding their suffering. It will take a sharp-eyed owner to notice that one of the flock isn't herself.

Always separate any chicken that appears unwell or is injured. Ideally the sick bird should be placed in a small coop within sight of the others so they don't pick on her when she returns. Alternatively the patient can be installed in a large box – make sure she is safe from predators and vermin.

Droppings

You can tell a lot about your chickens' health from their droppings. A standard dropping is brownish with a white tip (the urine). About one dropping in every ten will be sloppier and frothy. Occasionally a chicken will have a loose dropping or two, possibly as a result of something eaten, but if this continues it should be investigated, as should any other variations from the norm.

Appreciating the Causes of Disease

Factors that can contribute to poor health include the following, some of which have already been discussed in earlier chapters:

- Poor hygiene
- Overcrowding
- Incorrect feeding
- Lack of clean water
- Drinking dirty or contaminated water
- Lack of ventilation
- Internal worms
- Contact with vermin and wild birds
- Mixing young birds with adult chickens
- Dust and ammonia
- Infection brought from outside
- Stress
- External parasites

Contact with vermin and wild birds

Wild birds as well as vermin can bring disease and parasites into your flock. They are attracted by – and will contaminate – food and water, while sending your feed bills through the roof. Netting the top of the chicken run, or placing the feeder and drinker in a covered area, can help deter wild birds from joining the feast. Put everything away at night, and act quickly at the first sign of mice or rats (see Chapter 9).

Mixing young birds with adult chickens

Older chickens may carry low levels of some diseases to which they have built up immunity. Chicks and young birds don't have this protection, so are vulnerable to infection, as well as possible attacks from the older hens. Eating adult feed can also cause problems for young pullets (see Chapters 6 and 13).

Dust and ammonia

Chickens are particularly susceptible to respiratory diseases. Ammonia breaks down the defences that help prevent harmful organisms from gaining access to the lungs, air-sacs and hollow bones of the chicken.

Keep dust to a minimum, and don't leave cleaning the coop until it starts to smell of ammonia.

Infection brought from outside

A healthy flock can quickly be turned into a sickly one. Always keep new birds in isolation for at least three weeks. Don't let them come into contact with the existing flock, and be careful not to transfer disease yourself.

Many organisms can be communicated via human contact, clothing, boots and vehicles. Having an approved poultry disinfectant as a boot-dip outside the chicken run is a sensible precaution. If you visit poultry sales, shows or even neighbouring flocks, remember to disinfect boots and change clothes before tending to your own birds. Thoroughly clean and sanitize housing and equipment (including carrying crates) between occupants, and don't lend poultry equipment (or poultry) to others.

Stress

Chickens are easily affected by stress. When you consider the number of animals wanting to eat them, maybe they do have reason to be a little nervous.

Any of these could give your chickens cause for concern:

- Being moved
- Being handled if they are not used to it
- The addition of new birds
- Any sudden changes to their routine or feeding
- Possible predators
- Low-flying aircraft
- Rain or hailstorms
- Becoming soaked in a downpour
- Snow (both when it arrives and goes, due to the change of ground colour)
- Sudden loud noises
- Parasites

Stress makes chickens more vulnerable to infections. When stressful situations occur, they will benefit from a boost to their immune systems (e.g. apple cider vinegar or a vitamin supplement).

Combating External Parasites

You may have realized that there is a constant war between the poultry keeper and the various insects that torment chickens. Frequent checking of the birds and henhouse, thorough cleaning and the use of repellents are all weapons in this war, which has to be won if your chickens are to survive.

This may seem a bit extreme, but some parasites can kill chickens, or at least weaken them badly. Prevention is better than cure, because some of these bloodsuckers are very difficult to shift.

Let's start with the number one enemy:

Red mite

These are tiny grey/brown insects that turn red after they have fed on the chickens' blood. They live in the henhouse (rarely on the birds) and come out at night to feed. In warm weather they can reproduce in a week, and are able to survive in empty henhouses (i.e. without food) for months.

Red mite can transmit infections, but more commonly their attentions make the birds anaemic and weak. A bad infestation can result in fatalities.

Identifying red mite

Poultry red mite is a completely different insect to the red spider mites found in the garden.

Red mite aren't easy to see, although shining a torch into the henhouse at night may reveal them moving away from the light. During the day they hide in crevices, especially at the ends of perches. Running your thumb around a perch may result in a smear of red – that's a load of red mite you've just squashed!

The mites also leave heaps of grey dust, like cigarette ash, along ledges in the henhouse. You may even find red dots on the eggs – if the hens are still laying. An unexpected drop in egg production can be a sign of parasites.

Mites irritate the birds, making them restless. Unsurprisingly, you may find them reluctant to go into their house at night. If they have previously roosted there without trouble, then suspect red mite.

Red mite don't usually live on humans, but may bite them and can cause skin problems. You might start scratching after cleaning the henhouse, or even find your skin crawling!

Dealing with red mite

This is easier said than done – spotting their presence early and dealing with them promptly will give you an advantage.

If the henhouse has roofing felt, replace it with corrugated bitumen – red mite love to hide in felt, and if they are living in the roof everything else you do will be in vain.

Remove every scrap of bedding from the house (burn or seal it in bags as there will be mites in it), and take the house to pieces, if possible. Scrub it well, preferably with a product effective against red mite – get it into all the cracks to make sure it comes into contact with the mites. A watering can or sprayer can help ensure that all areas receive a soaking. The insects will start emerging to escape, so allow them to come out of hiding, then douse them again. A pressure washer is useful if there is a bad infestation.

When dry, use a red mite treatment on the house. Protect the birds by dusting them with a repellent too.

Unfortunately there are likely to be some mites or eggs remaining. It's almost impossible to get them all at the first attempt, and they will double their numbers within a week.

Therefore, you will need to take action a few days later in order to mop up any remaining mites. If it's impossible to scrub the henhouse again, a good spray with a suitable product may suffice. Liquid sprays are easier to get into all the crevices than powder (if using Poultry Shield for cleaning, keep back some diluted mixture in a sprayer – it can be used around the birds but not on them). The chickens should also have another dusting of repellent.

Red mite treatments

Diatomaceous earth is made of fossilized algae, and desiccates the outer coating of insects. It's a natural product that is usually suitable for applying to the birds as well as the house. It can be used as a preventative measure against parasites, and also to deal with an infestation.

When the house is clean and dry, generously apply diatomaceous earth, making sure you get it everywhere – it must come into contact with the mites. Rub it well into the perches, especially around the ends and undersides. As always, before using it on the birds, you should refer to the manufacturer's instructions.

There is a variety of other red mite treatments on the market, and more are appearing all the time. Some contain potent chemicals, but for a bad infestation it may be necessary to experiment with different products. You can buy smoke fumigation kits from poultry suppliers, and some people have found a wallpaper steamer very effective. A carefully used blow torch will do the job too – but obviously not in a plastic house!

At one time creosote proper (which is still available, but there are restrictions on its supply) was regularly used as a wood preserver, and red mite seem to have been less of a problem then. We found that applying creosote substitute and replacing the felted roof worked wonders (touch wood!). Creosote is fearsome stuff, but the substitute is less harmful. One of our hens managed to tip the paint kettle over herself – she needed a thorough wash, but is still going strong several years later. The mites didn't like it, though, and haven't come back. However, our henhouse can be taken apart, and every section was treated, inside and out. Other people have found that creosote substitute hasn't worked for them.

Whatever you decide to use, prompt action is essential. In an ideal world you would have a battery of products in stock, but if caught out use whatever can be sourced quickly rather than trying to obtain the latest super-duper mite killer. By the time it arrives, you will certainly need it!

Northern fowl mite

Red mite gets most of the attention, but this one is also a potential killer. It lives on the birds, and causes debility which can be fatal – look for it during your regular inspections.

Identifying northern fowl mite

These form greasy clumps of black, often around the bird's vent, but also on the head, particularly in ear canals and crests. Look closely and you may see the mites moving. They are brown/grey in colour.

Affected chickens will become restless and depressed. Combs and wattles will be pale. The birds will be scratching, possibly causing loss of feathers. They may start to drink more and laying will decline. You could also find the mites crawling up your arms after handling the birds.

Their breeding cycle can be less than seven days and numbers can increase in cool weather too.

Dealing with northern fowl mite

These are also difficult to eradicate, and the whole bird needs to be treated, not just the affected area. Again, prompt identification and treatment gives the best chance of success. Use one of the proprietary products designed to combat this mite, dusting it all over, including a small amount in the ear canals. Repeat the treatment within seven days.

Even if only one or two birds are obviously affected, you should give the whole flock a preventative dusting and scrub the coop too.

Scaly leg mite

This burrows into the skin under the leg scales, causing irritation and discomfort to the bird. If not treated, it will cause lameness and even loss of toes.

Identifying scaly leg mite

The leg scales will be raised and thickened, with white crusts underneath – this is the excrement of the mites. If left untreated, the legs will start to look lumpy and the bird may have trouble walking. The mite lives on the birds and is spread by contact.

Dealing with scaly leg mite

Trying to pull off the crusts will cause bleeding. Soak the legs in warm, soapy water, and clean them with a soft toothbrush. Dry well and dip each leg into a wide jar of surgical spirit (as long as there is no broken skin). Cover the legs with petroleum jelly to smother any remaining mites. The treatment will need repeating weekly for three to four weeks. Clean the house thoroughly too, and apply an anti-mite product.

Alternatively, you can buy products that relieve the itching and kill the mites (creams are easier to apply than sprays). These need to be reapplied regularly, according to the manufacturer's directions.

The leg scales will be shed at moulting, so you may not see much improvement until the next moult.

Lice

Lice cause the chickens to scratch, preen excessively and lose feathers. A heavy infestation will be debilitating, and egg production will drop.

Identifying lice

You may find these small grey insects on the skin, but they quickly move away from the light. It's easier to see their eggs – white clumps stuck to the base of the feathers, usually around the vent.

Dealing with lice

Adding diatomaceous earth to the dust-baths, bedding and nest-boxes helps chickens to keep lice under control, but if there is an infestation you should dust the chickens with one of the proprietary products.

Lice eggs take up to three weeks to hatch, so you should continue the treatment at weekly intervals, as recommended by the manufacturer. Badly encrusted feathers can be pulled out and burnt.

Other external parasites
Fleas

Chickens can also get fleas, which are usually found as dark brown spots in the head feathers. Much of the flea's life cycle takes place in the house or bedding, so this will need to be treated as well as the hens.

Ticks
Free-range chickens may pick up ticks from long grass. Trying to pull them off results in the heads being left behind, which can cause infection. Dabbing them with surgical spirit or smothering them with petroleum jelly will cause them to drop off.

Depluming mite
This is a relative of the scaly leg mite. It burrows into feather shafts and causes the chicken to pull out its own feathers. These mites are very small and difficult to see. Unusually they produce live young rather than eggs and take three weeks to complete their life cycle. They are spread by direct contact.

Ivermectin
This drug is sometimes used to deal with external and internal parasites. It should only be used on chickens in consultation with your vet.

Although Ivermectin can be purchased under the Small Animal Exemption Scheme for non-food producing animals, it isn't licensed (as it hasn't undergone full clinical trials) for chickens. Vets may prescribe it using their clinical judgement, but there will be a withdrawal period when eggs cannot be eaten. Once a hen has been treated with Ivermectin her eggs should never be offered for sale.

Recognizing Infectious Diseases
This is only intended as a brief guide to some of the most common or important diseases. Although it's useful to know the symptoms, only a vet can give an accurate diagnosis and prescribe antibiotics if necessary. If several birds are affected by a condition, or there are a number of unexplained deaths, veterinary advice should be sought without delay.

Mycoplasma
This very infectious respiratory disease can appear within a few days of being contracted. Symptoms are foamy eyes, swollen sinuses, rasping breath and sneezing. There is often a sickly smell. Another strain of the disease produces hot and swollen joints.

Mycoplasma can prove fatal to some chickens, while others in the flock may show few signs. Antibiotics administered in the initial stages can help aid recovery.

Birds that recover from an infection are liable to further attacks and will remain carriers of the disease. If young stock is introduced to a flock that carries mycoplasma they are likely to become infected, even though the older birds remain apparently healthy.

Mycoplasma can be transmitted by poultry or wild birds. Stress makes chickens more vulnerable to this infection, so any that have been in poultry sales or shows should be quarantined and carefully monitored. The disease is able to survive for several hours on clothing and poultry equipment. It can also be present in hatching eggs. There is a vaccine.

There are several types of respiratory disease, and more than one can be present – a blood test may be required to determine the cause or causes of the symptoms.

Infectious bronchitis

Signs are similar to mycoplasma, but the disease spreads rapidly through the entire flock. Following the respiratory symptoms, there may be kidney damage and infection of the reproductive organs. This will be revealed in soft-shelled or misshapen eggs, and the bird will remain a carrier of the disease.

Damage to the oviduct can also result in eggs descending into the abdomen, leading to egg peritonitis.

The virus is killed by disinfectant, so good hygiene and keeping stress to a minimum are effective preventative measures. There is a vaccine, which must be given when the birds are young.

Aspergillosis

In humans this disease is known as 'farmer's lung', and is caused by fungal spores in mouldy hay, straw or bark. Signs are breathlessness, lethargy and thirst, but the disease shows few symptoms until it is advanced. Successful treatment is difficult.

Coccidiosis

Coccidia eggs (oocysts) are picked up by the bird and develop in the gut. They multiply rapidly and feed in the intestine, causing destruction of the gut walls. Eventually more oocysts are passed in droppings to continue the cycle.

The disease often proves fatal, especially in young birds. Older chickens (apart from ex-battery hens) have frequently been exposed to low levels of infection, giving them some immunity, but can still be at risk in certain conditions.

The birds stop eating, and become hunched and miserable, with ruffled feathers. There is often watery diarrhoea, which may be white or have blood in it. Immediate veterinary attention will be required.

The vet can prescribe treatments, but prevention is better than cure. Traditionally, chicks have been given feed containing an anti-coccidiostat, but a vaccine is now available (see Chapter 13). Check when buying young birds that they have been protected against coccidiosis.

Good husbandry is essential too. The oocysts can be transmitted on boots and clothing, and are very resilient to the usual disinfectants. They are killed by extremes of temperature, but can survive for long periods in empty housing.

Warm, humid conditions (e.g. a brooder or shed with damp bedding) are ideal for the oocysts to enter the first stage of their development. When eaten by a chicken, they will be ready to start multiplying.

Indoor birds are most at risk – bedding must be kept clean and dry. Standing drinkers on wire mesh helps to keep the birds out of wet floor litter. Make sure feeders and drinkers are free from droppings and cleaned regularly.

Good ventilation helps to avoid humidity. Stress factors, such as overcrowding, will encourage disease.

Use a product designed to kill coccidia oocysts before moving new chickens into housing previously occupied by other birds. The ground is also liable to contamination, and should be treated with a suitable sanitizing agent if birds can't be regularly moved.

Marek's disease
This herpes virus is usually seen in young birds, especially pullets coming into lay. Stress is particularly likely to cause the virus to surface. The usual signs are birds becoming thin and weak, with lameness leading to paralysis of legs and wings. Sometimes birds die suddenly without any obvious symptoms. Affected birds should be culled – there is no cure and any birds that do survive will remain carriers.

The virus is mainly passed on via feather debris, and can survive for at least twelve months in henhouse dust. Some breeds are particularly susceptible to this disease (e.g. Silkies and Sebrights), but chicks can be vaccinated.

Avian influenza (bird flu)
This is a notifiable disease, which means that DEFRA must be informed of any occurrences. As well as causing numerous deaths in birds, there is also a potential risk of the virus creating a strain that can spread to humans. This makes avian influenza important enough to make the national news whenever

there is an outbreak. Unfortunately it has reached Europe, and there have been some cases in the UK.

The disease is transmitted through direct contact, or by body fluids and droppings. It can be spread via infected poultry, wild birds (particularly waterfowl, which can carry the disease while showing few signs), vermin and human transference. The virus can survive for around fifty days, or even longer in wet conditions.

There are two types of the virus, but Highly Pathogenic Avian Influenza (HPAI) is usually the most serious. Symptoms vary from mild to severe, and include loss of appetite, swollen heads, respiratory distress, blue discolouration of neck and throat, diarrhoea, decline in laying – and deaths. Should there be an outbreak in a flock, any birds that survive will be culled and restriction zones put in place around the area.

Not only do infected birds suffer and die, but there could also be significant effects on the poultry industry. As mentioned in Chapter 3, it would be wise to be prepared for controls to be imposed whenever necessary to try and exclude avian influenza from the UK. Keep an eye on the DEFRA website and poultry magazines (see Further Reference) for updates on the situation and ways to protect your chickens. If free-range chickens suddenly have to be confined, remember to provide some 'entertainment' to prevent boredom leading to squabbles (see Chapters 5 and 7 for ideas).

In general, your best defence against avian influenza (and other diseases) is good hygiene, keeping food and water away from wild birds, vermin control and close observation of your chickens.

You should immediately inform your vet if you suspect avian influenza, or there are unexplained deaths in the flock.

Newcastle disease

Although rare in the UK, this is included because it's a notifiable disease – an outbreak must be reported to DEFRA. Apart from being fatal to chickens, it can cause flu-type symptoms and conjunctivitis in humans.

The most obvious indication is a large number of deaths in the flock. Other signs are varied and include respiratory symptoms, severe diarrhoea or paralysis where the neck twists. Laying may suddenly decline, with a high proportion of soft-shelled eggs.

If you think your birds may have contracted Newcastle disease, or experience many unexplained deaths, you should contact your vet immediately.

The disease can be carried by wild birds, especially pigeons, and also on clothing. There are disinfectants that will destroy the virus, and there is a vaccine.

Salmonella

This rarely causes illness in chickens, but can cause severe diarrhoea and vomiting if passed to humans. Commercial flocks are routinely vaccinated and tested. Small flocks can also be tested by means of a droppings sample. Maintaining good hygiene in the kitchen will minimize risks.

Chicken Pox

This is nothing to do with chickens! Most likely the name came from the Old English word 'gican', meaning itching – itching pox.

The unrelated fowl pox affects chickens only and causes raised, crusted areas on the skin. It's painful and makes the birds miserable, but isn't usually fatal unless it infects the windpipe and oesophagus.

Heart Disease

Chickens can suffer from heart attacks, especially overweight heavy breeds. A purple comb is often an indication of circulatory problems. If a bird starts turning purple in the face, comb and wattles when handled, replace it gently on the ground immediately, as this is a warning of an imminent heart attack. Try to avoid handling the bird in future, and keep it free from stress.

Respiratory System

Chickens have only small lungs, which don't expand like those of mammals. The air is pushed around their bodies through a series of air-sacs. It's a very efficient breathing system, but can allow airborne infections to penetrate deep into the body and bones. The air-sacs can be punctured by deep wounds, and holding birds too tightly can result in suffocation.

Digestive Problems

When a chicken swallows food, it is stored in the crop at the base of the neck. After the chicken has eaten, the full crop can easily be felt.

There are a couple of possible problems that can occur in the crop.

Sour crop

This yeast infection can be caused by unsuitable or mouldy feed, a deficiency of vitamin A (found in green vegetables), or by oral antibiotics.

The bird may be sluggish, but mostly you will notice the smell from its mouth. The vet should be consulted.

Apple cider vinegar in the drinking water one week each month is a preventative measure (20ml of vinegar to a litre of water in a plastic drinker).

Impacted crop

A chicken's crop is quite large after it has eaten, but should be empty by the morning. If it is still bulging and feels firm, there may be a blockage. This is often caused by the chicken eating long-stemmed grasses or swallowing something unsuitable like string. The mass will prevent the bird from eating and she will starve if left untreated.

There are various remedies used by poultry keepers to clear an impacted crop, but the novice chicken owner should seek advice from the vet. Unfortunately the crop muscles may be weakened by the impaction, causing the situation to recur.

Keep grass short to help prevent this condition.

Egg-Laying Problems

Pullets coming into lay and hens at the end of their laying lives are more likely to experience difficulties. Ex-battery hens are often susceptible, as are young pullets that have started laying too early.

Egg-bound hens

An egg-bound hen is unable to push out her egg. Sometimes this happens when the egg is too big, but it can also be caused by a calcium deficiency – liquid calcium may help an egg-bound hen.

The hen will go in and out of the nest-box, look miserable and stop eating. You might see her straining to pass the egg and will possibly be able to feel it in her abdomen.

Separate the hen from the others. Make her as warm and comfortable as possible – a hot water bottle may help. If you need to check for the egg, use a well-washed lubricated finger or wear disposable gloves to avoid introducing infection.

Try smearing a little lubricating jelly around the inside of the vent and massaging the abdomen gently – you could use a warm flannel. This may be enough to do the trick, but be careful the egg doesn't break inside the hen.

Other ideas include sitting the hen in a bath of warm water for twenty minutes (don't let it cool down) or holding her over a bowl of steaming water to warm the vent and abdomen. You could drape a towel over her rear end to concentrate the steam – make sure she's not close enough for the steam to scald her.

If the hen still cannot pass the egg, it will have to be punctured to release the contents and every piece of shell removed. The sharp edges of shell could damage the oviduct and cause infection so at this stage you may wish to seek professional help.

Let the hen recover in a warm place, away from the flock. There may have been muscle damage, which could lead to a prolapse of the oviduct.

Prolapse

A prolapse is when the oviduct protrudes through the vent. Unfortunately other chickens will find this irresistible, and if not spotted in time the hen may be killed by her companions.

A small prolapse can sometimes be pushed back in, using a clean (gloved) finger and some lubrication. If a large amount of tissue is protruding, the vet should be consulted and the hen will probably need antibiotics.

There is a possibility of the problem recurring when the next egg is laid. A diet of only grain may slow down egg production, giving the prolapse a chance to heal.

It has been found that the suprelorin implant, used for controlling fertility in male dogs, temporarily suspends laying in hens. The implant can only be administered by a vet, lasts about three months, and is fairly expensive. It is even possible for a hen to be spayed, although this is costly, and she might not survive the operation.

It is usually kinder to cull a hen who continuously suffers from prolapse, as she will always be liable to painful attacks from the other chickens.

Egg peritonitis

This occurs when an egg yolk goes into the abdomen instead of down the oviduct. It may be due to a damaged oviduct or the result of stress.

Sometimes one yolk can be dealt with by the body, but an infection may set in, especially if several yolks descend into the abdomen. The bird will be in pain and dejected – the abdomen may be swollen. The condition is difficult to diagnose and likely to be fatal, although antibiotics may sometimes help if it is caught early enough.

Treating Wounds and External Problems

Clean small cuts, and apply antiseptic spray. Separate the bird until the wound has healed. More extensive damage will need veterinary attention.

Bumble foot

This is caused by an infected wound under the foot. It can be the result of a heavy bird jumping down from a high perch or because of a cut. A pus-filled swelling develops and the bird becomes lame. It may be possible to squeeze out the pus, in which case clean the wound and apply antiseptic. Cover with a dressing and keep the bird on dry bedding, changing the dressing frequently. Severe cases will need antibiotics for the infection, which can become serious if left untreated.

Convict's foot and toe balls

Wet bedding combined with mud and droppings can set like concrete around the feet and toes. Trying to pull this off will damage the foot. Soak with warm, soapy water to loosen it, clean and dry the feet – and replace the floor litter!

Frozen combs and wattles

Breeds with large combs, especially cockerels, can sometimes suffer from frostbite. Good ventilation in the henhouse helps prevent this, but if the weather is particularly cold, rub petroleum jelly into the combs and wattles each evening. If frostbite does occur, treat with antiseptic spray – bad cases may require the vet.

Understanding Feather Loss

Moulting

This can be alarming the first time it happens. You open the henhouse door one morning and find feathers all over the floor – it looks as if the fox has paid a visit!

Moulting usually occurs at the end of summer or beginning of autumn, depending on when the bird was hatched. Pullets lay for up to a year before moulting and then it becomes an annual event. While some lose only a few feathers, others can look as if they have been through a particularly fierce wind tunnel. A once fluffy hen may resemble something that's just been released from a battery cage.

Face, wattles and comb become paler – the comb shrinks too. Chickens are under some strain at this time, and as they concentrate their resources into growing new feathers, egg-laying gradually decreases and stops.

The moult may last a few weeks (older hens take longer), and some hens won't lay again until spring. Feathers are mainly made up of protein, so feeding plenty of good-quality food (keep grain and treats to a minimum to make sure they eat it) will help chickens re-feather. Some people like to give their birds a tonic to help them through the moult.

Eventually small 'pin' feathers will start to show. These are full of blood, so the chicken may be sensitive to being handled at this time.

Clipped wing feathers will grow back too, but don't re-clip them until the quills are white again (see Chapter 8).

Check for parasites if the birds don't re-feather as expected.

Other reasons for feather loss

Some birds go through a 'partial moult' and might lose just a few neck feathers. Stress or parasites can also result in loss of feathers.

Chickens sometimes eat feathers to compensate for lack of protein in their diet, and may pull out each other's feathers – 'feather pecking'.

Boredom, overcrowding or parasites are also causes of feather pecking, which can turn into a serious problem. Eventually blood will be drawn, leading to further attacks and even fatalities.

You can use an anti-peck spray, which is designed to make the feathers taste unpleasant to the birds, but you should also eliminate the cause of the problem. Make sure the chickens have enough space, and consider whether the mixture of breeds is appropriate (some breeds are more easily picked on, while those with muffling and crests often become targets). Cut out grain, and make sure the chickens are eating the correct diet. Let them free-range if possible, and supply plenty of distractions in the run. If there is a main perpetrator, put her where she can see the others but not pick on them.

Watch out for feather pecking during the moult, as the partially feathered birds will be particularly vulnerable.

Hens sometimes have bald patches on their backs due to an over-attentive cockerel. They should be separated from him, or else their backs can be protected by poultry saddles. These provide an ideal hiding place for parasites, so the hens should be sprayed with repellent and inspected frequently.

Looking After Yourself!

In order to look after your chickens you need to stay healthy too. There are a few common sense precautions you can take:

- Always wash your hands after dealing with chickens
- Wear gloves – either heavy duty or disposable – whenever appropriate
- Use overalls or old clothes for messy jobs, and change out of them before sitting down indoors
- Always use a face-mask when spreading powders (even those safe for chickens)

- Keep all medication and cleaning products out of reach of children and pets
- Dispose of any unused medication and packaging properly (the vet can help with this)
- Only use drugs that are licensed for chickens, unless instructed by the vet
- Maintain good hygiene standards when dealing with eggs and chicken meat

Key Points
- Seek veterinary advice if concerned about your chickens' health
- Check your birds regularly and separate any chicken that is unwell or injured
- Standard droppings are brown with a white tip – around one in ten may be sloppier and frothy
- Good hygiene and care will help prevent disease
- New birds should be isolated for at least three weeks
- Diseases can be transferred via humans
- Stress makes chickens more susceptible to infections
- External parasites can weaken chickens and cause fatalities
- Red mite live in the henhouse and feed on chickens at night
- Northern fowl mite live on the birds and look like greasy black clumps
- Scaly leg mite causes scales to become raised with white crusts
- Lice are small grey insects – the eggs are white clusters at the base of feathers
- Chicken fleas are usually found as brown spots on the head
- Ticks should be dabbed with surgical spirit or smothered in Vaseline
- Depluming mite burrow into feather shafts, causing the chicken to pull out feathers
- Ivermectin should only be used in consultation with the vet, requires an egg-withdrawal period, and subsequently eggs should never be offered for sale
- If several birds are unwell, or there are many unexplained deaths, contact the vet at once
- Symptoms of mycoplasma are foamy eyes, swollen sinuses, rasping breath and sneezing
- Infectious bronchitis has similar symptoms to mycoplasma but spreads rapidly and can cause reproductive disease
- Aspergillosis is caused by fungal spores in mouldy organic matter

- Humidity and poor hygiene are ideal conditions for coccidiosis
- Marek's disease usually causes paralysis of legs and wings
- If you suspect avian influenza or Newcastle disease contact the vet immediately (DEFRA must be notified of an outbreak of either disease)
- Chickens can carry salmonella, which causes diarrhoea and vomiting in humans
- A purple comb indicates circulatory problems and chickens can suffer heart attacks
- Air is pushed around the chicken's body through a series of air-sacs
- Sour crop is a yeast infection causing bad-smelling breath and listlessness
- The crop may become blocked (impacted) by long grass or foreign objects
- An egg-bound hen is unable to push out her egg
- A prolapse is when the oviduct protrudes from the vent
- Egg peritonitis occurs when a yolk goes into the abdomen and is usually fatal
- Bumble foot is a pus-filled swelling on the base of the foot
- Chickens moult annually
- Poor nutrition, boredom, overcrowding or parasites can cause feather pecking
- Take sensible precautions to safeguard your own health when keeping chickens

KIDS' CORNER

Quiz

How much do you know about keeping your chickens healthy?

QUESTION ONE

What's the first thing you should you do if you think a chicken is ill?

   (a) Leave her in the henhouse to rest for a while

   (b) Separate her from the other chickens

   (c) Put her in another coop with a couple of her friends for company

QUESTION TWO

Which of these statements is wrong?

   (a) Chickens cope well with stress

   (b) Keeping the henhouse clean helps prevent disease

   (c) Chicken diseases can be carried on human boots and clothing

QUESTION THREE

Where do red mite usually live?

   (a) On the chickens

   (b) In the chicken house

   (c) Under the ground

QUESTION FOUR

What is the crop?

   (a) The pouch at the base of the neck where food is stored after being swallowed

   (b) The muscular stomach where food is ground

   (c) Another name for the chicken's intestines

QUESTION FIVE

When chickens moult will they:

   (a) Start laying eggs for the first time?

   (b) Carry on laying as usual?

   (c) Stop laying eggs?

ANSWERS

One (b); Two (a); Three (b); Four (a); Five (c)

Well done if you knew all the answers!

Chicken Chat

'You can't make an omelette without breaking eggs': This saying suggests that it may be necessary to cause some damage in order to accomplish a result. For example: 'Emily was in trouble for having wet clothes after scrubbing the henhouse to destroy the red mite – she told her mum that you can't make an omelette without breaking eggs'.

Chicken Jokes

What did the hen say when she laid a square egg?
'Ouch!'

What was wrong with the sick chicken?
She had people-pox!

Something to Do . . .

Look at your chickens every day. Watch them carefully and notice how they behave. If you know how they look when they are well, you will quickly notice when something is not quite right. Spotting a problem early gives you the best chance of helping the chicken to get better.

CHAPTER TWELVE

Having Something to Crow About: Cockerels

If you have neighbours, keeping a cockerel is very likely to cause problems. All cockerels crow – even little ones can produce a piercing shriek.

Keeping the chickens shut in until a reasonable hour in the morning will help. A solid henhouse muffles the noise, and the cockerel may not crow if the house is dark.

Another tactic is to shut the cockerel in a box at night and put him where he can't be heard. The box should be well ventilated and large enough for him to perch comfortably.

Cockerels don't just crow at dawn (which comes early in summer when bedroom windows are likely to be open). Some cockerels crow at night, especially if they are disturbed. Throughout the day, cockerels will shout a challenge to other males, announce they have done their duty with a hen, sound the all-clear – or simply crow because they enjoy it. Neighbours working night shifts are unlikely to be amused.

Even rural poultry keepers have experienced protests about cockerels, and the Environmental Health Department will investigate complaints.

Aggression in Cockerels

Some cockerels are protective of their hens or feed, and become aggressive towards people. Don't underestimate a hostile cockerel – spurs can shred jeans, even wellies, and may also cause a nasty infection.

Even little cockerels can be feisty, and are often good flyers, bringing them dangerously close to face level.

Obviously such birds must be kept well away from children, and are unlikely to be a successful addition to the garden.

Although some breeds are more likely to be aggressive, it isn't possible to guarantee that any cockerel will be placid. Some change temperament with maturity or during the breeding season – tame cockerels can be particularly difficult, as they have no fear of humans.

Various remedies have been suggested for cockerels that play dirty, but their long-term effectiveness is debatable. I've found a large water pistol quite handy, but bearing arms isn't always practical. A dustbin lid also makes a good shield, but really there is only one solution to a troublesome cockerel – which is probably why they invented Coq-au-Vin.

Are all cockerels aggressive?

Not all cockerels are grumpy, and some are delightful birds. We inherited three in our first flock and they gave no trouble. They were handsome chaps, much admired, and all got along (reasonably) well. Keeping several males isn't usually advisable as they are likely to fight.

Deciding Whether You Need a Cockerel

Are you sure you need a cockerel? Hens won't lay any better because there's a male around. His attentions may even cause them distress or injury, especially if there aren't enough hens to keep him busy or if he singles one out as a favourite.

If you want chicks, fertile eggs are easily bought and you can hatch any breed you fancy.

It's usually better to begin by keeping hens alone, and think about adding a cockerel later when you have settled into chicken life.

Advantages of keeping a cockerel

In a free-range flock a cockerel will act as lookout for the hens and sort out squabbles. I've noticed that my hens seem more settled with a cockerel to look after them.

He will also fertilize the eggs (there is no obvious difference in appearance or taste), allowing you to breed replacements. Bear in mind that at least half the chicks are likely to be . . . more cockerels.

Hatching Cockerels

This is often overlooked in the excitement of putting eggs to hatch, but the proportion of males to females is usually quite high.

If you have an 'auto-sexing' breed, the males and females will have different-coloured down. You can also cross two different coloured birds to produce chicks that can be sexed at a day old (e.g. a Rhode Island Red male crossed with a Light Sussex hen will give light males and darker females). If you don't want cockerels, this will mean culling the chicks.

Otherwise, it can be difficult to identify the cockerels until the chicks are well feathered – sometimes you have to wait until they start crowing!

This may give their owner a headache in more ways than one. If a single cockerel can cause unrest, half a dozen could result in a neighbourhood uprising. Even if this isn't a problem, keeping them all is unlikely to be practical. Apart from fighting, they will harass the hens (who may stop laying in protest) and eat vast amounts of food.

We kept a young cockerel who threw his own father out of the flock when he reached maturity. Unfortunately for Junior, he was still young enough to become a very delicious chicken dinner – but his poor old dad never quite recovered from the humiliation.

Re-homing cockerels

This isn't easy. Many people aren't able or don't want to keep one, while those who do have a cockerel aren't usually in the market for more.

Pure-breeds (especially if they are rare) may have a better chance of finding a home than those of mixed parentage. It's worth trying a few adverts on local notice boards or on poultry websites. Some (very few) rescue centres might take unwanted poultry.

Make sure your cockerels go to caring owners, and be wary about offering them 'free to a good home'. Illegal cock-fighting organizations apparently use tame cockerels to give confidence to young fighting birds. It would be better to cull the unwanted males than abandon them to such a fate.

Producing Table Birds

Surplus cockerels can also be grown into meat birds, but hatch a suitable breed to produce the best results. Light laying types will never turn into full-breasted roasting birds, no matter how much feed they eat. Take into account that traditional chickens may require several months to reach a respectable weight, rather than the six weeks allotted to commercial meat birds.

Keep the young cockerels separately so they don't bother the hens. If they squabble, the introduction of an adult male will often restore calm.

Preparation of table birds

There are many regulations involved in selling meat birds, but you can prepare and eat your own.

The practical skills required are better learnt from an experienced person than a book. Chickens should be despatched cleanly and confidently – the usual method at home is neck dislocation. There are various gadgets which are supposed to make the process easier, but there have been horror stories about them not working properly.

You can attend courses that will take you through the entire process from despatching to trussing. Alternatively, an experienced poultry keeper (or gamekeeper) could show you what to do. It takes a while to pluck a chicken, although it becomes quicker with practice. The bird should then be hung for a few days (depending on the weather) before being eviscerated. This needn't be a messy job but, having learnt with a knife in one hand and a book in the other, I can really recommend having a practical lesson first.

A free-range cockerel that you have raised and prepared yourself is a very special meal. Don't expect the meat to be white and bland – it may be dark, rich and even a little gamey. There could also be a slight tendency to dryness, so baste it with butter or olive oil. Pot roasting works particularly well.

From egg to table takes time and patience – but will result in a taste of a bygone luxury.

Key Points

- All cockerels crow, and even rural neighbours may complain
- Not letting cockerels out too early helps reduce morning disturbance
- Cockerels crow throughout the day and sometimes at night
- The Environmental Health Department will investigate complaints
- Some cockerels are aggressive towards people and can cause injury
- A cockerel's temperament may change with maturity
- There are few advantages to keeping a cockerel, and it's usually better to start with just hens
- Hatching eggs will result in several males
- Sometimes it's difficult to identify the cockerels until they start crowing
- Cockerels often fight each other
- Re-homing cockerels is difficult – be careful they go to good homes
- Choose a suitable breed if intending to produce table birds
- Despatching and preparation of table birds should be learnt from an experienced person

KIDS' CORNER

Quiz

What have you found out about cockerels from Chapter 12?

QUESTION ONE

How noisy are cockerels?

(a) They crow at any time of the day or night

(b) They only crow in the morning

(c) Not all cockerels crow

QUESTION TWO

What wouldn't you expect a cockerel to do?

(a) Encourage the hens to lay more eggs

(b) Fertilize the eggs

(c) Look after the hens

QUESTION THREE

Which of these statements is wrong?

(a) A cockerel may become more aggressive as he gets older

(b) Keeping more than one cockerel is likely to result in fights

(c) Cockerels never attack humans

QUESTION FOUR

If you hatch some chicks, what are you likely to get?

(a) Only females

(h) Mainly females

(c) More males than females

QUESTION FIVE

If you have hatched some spare cockerels, which statement is likely to be true?

(a) It will be easy to sell them to other chicken keepers

(b) It will be difficult to find suitable homes for them

(c) It's a good idea to keep them with the rest of the flock

ANSWERS

One (a); Two (a); Three (c); Four (c); Five (b)

How did you cock-a-doodle-do in the cockerel quiz? If you got them all right, you'll have something to crow about!

Chicken Chat

'Having something to crow about': When a cockerel is feeling pleased with himself, he crows to tell the world how clever he is. This saying is used when someone has reason to be proud of an achievement. For example: 'Oliver was hopeless at football but he trained really hard and now he's scored the winning goal – he certainly has something to crow about!'

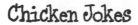

Chicken Jokes

Why did the cockerel cross the road?
He wanted to impress the chicks!

If a cockerel laid an egg on top of the henhouse, which way would it roll?
Cockerels don't lay eggs!

Something to Do . . .

You know that cockerels can make a lot of noise – but hens aren't silent either. Have you noticed how many different sounds your hens make? Do you know if they are telling each other about food, or danger, or that an egg has been laid? Listen carefully and see if you can work out their special language.

Can you copy their noises? If you can make the 'I've found food' sound, they will come running whenever you call!

CHAPTER THIRTEEN

Don't Count Your Chickens Before They've Hatched: Managing Broody Hens and Hatching Eggs

Hatching eggs and rearing chicks takes some skill and experience – but if you have a broody hen she'll show you how it's done!

Recognizing a Broody Hen

Not all hens go broody. Hybrids and many of the light laying breeds very rarely start sitting, and if they do they may give up or reject the chicks. Even so, chickens can be full of surprises (see 'Chickens Don't Always Follow the Rules' in Chapter 4).

A broody hen will sit firmly in the nest-box and show no interest in food. This could be a sign of illness, but the key is her puffed-up feathers. Approach her and she will squawk crossly or make a low growling sound. She may try to peck you, but a broody is usually so intent on sitting that she has no energy for anything else.

Lift her out of the nest-box (wear gloves if she objects strongly). Her breast will feel extra-warm, and she might have pulled a few feathers out. Place her on the ground. She may sit there in a daze for a few moments, then suddenly come to life and rush off noisily to eat.

A broody hen (as opposed to a sick one) will be keen to eat once awakened from her stupor. She may peck around for a bit but will soon be back on the nest. If you are hoping to hatch some eggs, this is good news – otherwise you will have to dissuade her from sitting.

Dealing with Unwanted Broodiness

An unwanted broody shouldn't be left to sit – it's not good for her, the other hens or your egg supplies.

A broody hen only eats, drinks and does a dropping once a day – it's hard work. Even if you remove the eggs, she may stay broody for several weeks. This will weaken her; she could even die. Giving a broody some imitation eggs so she can fulfil her instincts is not a kindness.

The broody will stop laying, and may also prevent other hens from using the nest-boxes. The resulting scuffles can result in broken eggs and hens laying elsewhere. The broody will sit on any eggs she sees, keeping them undesirably warm, and starting development in fertile eggs.

This may be annoying but a broody hen is only following her instincts and should be treated kindly – but firmly.

Discouraging a broody hen

Collecting eggs regularly helps discourage broodiness. Some hens will stop sitting if they are repeatedly removed from the nest-box or shut out of the henhouse for a while.

This can be inconvenient and is unlikely to be successful with determined broodies. The alternative is to put the hen in a separate coop where she can't sit.

An anti-broody coop

This is a coop with a wire floor, placed on legs or bricks so that the air can circulate underneath. The idea is to cool the hen and prevent her from settling comfortably. You can use a wire dog-crate – take out the floor panel and turn the crate upside down if the mesh on the bottom is too wide.

The broody must stay in this coop for several days, so make sure she has shelter from the elements. She must also be safe from predators and vermin (especially if using a dog-crate, which offers no protection). If keeping her indoors lay newspaper under the coop – broody hens do some very unpleasant droppings! Don't forget food and water.

Give layers' feed (no corn) to encourage the hen back into lay. You should soon see a change in her – she will be up and about, wondering why she's in this strange house. Try her back in the flock and see what she does – if she rushes back to the nest-box, she obviously needs more time in the 'cooler'.

This really works, and is much easier than battling with a determined broody all summer. The hen will only be inconvenienced for a short while, and will then be able to enjoy the sunshine with her friends again.

Secret broodies

If you have breeds that are likely to go broody, be suspicious if one unexpectedly stops laying. Watch her closely or she may disappear one evening, and it will take luck and a good torch to find her – sitting on a big pile of eggs under a bush. We found our Pekin deep in a woodpile one night, just before the light faded.

Occasionally a lost hen comes strolling back with a clutch of chicks in tow, but a hen sitting outside is vulnerable to all kinds of predators.

Deciding to Hatch Some Eggs

Sometimes owners start feeling broody too – what could be more delightful than the patter of tiny claws?

Watching a hen hatch and raise her chicks is fascinating. She will instinctively know what to do and will give her chicks an excellent start, teaching them all the ways of the chicken world.

When the chicks are teenagers the hen will return to the flock, leaving the owners to deal with her offspring (chickens really are very intelligent). As mentioned in the previous chapter, there are likely to be several cockerels amongst the chicks.

Consider what will happen to the males before deciding to hatch some eggs. One way around the problem is to buy hatching eggs from a breeder who is prepared to take back the cockerels. The birds will probably be despatched, but at least you will know they have been treated responsibly.

Vaccinations and Coccidiosis

You may wish to consider vaccinating the chicks, especially if you have a breed that's susceptible to a certain disease (Marek's in Silkies, for example). Many vaccinations are only effective if given at a very early age, so speak to your vet before the chicks hatch.

You can vaccinate against coccidiosis, or buy chick crumbs and then growers' feed medicated with an anti-coccidiostat (ACS) – don't do both, as one cancels out the other. Keep the medicated feed away from other animals, especially waterfowl. If it is eaten by laying hens there is usually an egg-withdrawal period – bear this in mind if a broody hen comes back into lay after sharing the feed with her chicks.

Hatching with a Broody Hen

Choosing a broody

A broody hen will hatch any eggs – they need not even be chicken eggs. Select a breed with strong maternal tendencies (such as a Silkie, Pekin or Wyandotte). Consider also the size of the hen – bantams make great mothers but can't cover the same number of eggs as an Orpington or Cochin.

It's unfair to ask a hen to hatch eggs when she's already been broody for some weeks – she'll be exhausted and may even give up halfway through.

Occasionally a pullet goes broody, but she shouldn't be allowed to sit unless she's been laying regularly for at least several weeks.

Bear in mind that even the broodiest hen can go off the idea, especially if she's never raised chicks before. Sometimes eggs can be given to another broody, but it can be more of a headache if a hen abandons her newly-hatched chicks. Read the sections on artificial incubation and brooding chicks, just in case you need to give nature a helping hand.

Encouraging broodiness

While removing eggs can discourage broodiness, leaving pot eggs in the nest-box can turn a hen's thoughts to motherhood – although some breeds go broody at the sight of an egg anyway.

Spring chicks

Most hens go broody in spring and early summer, but some try again later in the year. This isn't a good time to start a hatch. Naturally-raised chicks need the long warm days of summer to grow and thrive (when planning summer holidays, remember you will have an extra pen of birds to deal with).

Setting up a nursery

The nest-box isn't a suitable environment for hatching eggs. The broody will find it difficult to do her job properly while competing for space with the laying hens. Any chicks that hatch will be at risk from disease and parasites, as well as attacks by the adults.

A broody hen needs a separate coop, where she can hatch her eggs and raise her chicks in peace.

Select the broody coop with care – remember that little chicks won't manage a steep ramp – and make sure there is adequate ventilation and security. It must be proof against rats (ours is completely wrapped in a double layer of wire). Chicks grow rapidly, so the run should be large enough to accommodate the growing family, or capable of being easily extended.

This is ideal as a broody coop, and can easily be moved to fresh ground

Making a nest

A piece of turf makes an ideal base for a nest, and provides the necessary humidity. In very dry weather it can even be watered to keep it moist. Scrape away some of the soil under the turf so the grass forms a slight dip in the middle, then construct a comfortable nest of short straw or shavings, with enough space for the eggs to form a level circle. Place a couple of pot eggs in the nest.

Treat both coop and hen against parasites, and then move her at night with as little fuss as possible. The move may disturb her, and she might even change her mind about being broody, but if she's really committed she will quickly settle on the dummy eggs.

Adding the eggs

When the broody has settled, you can replace the pot eggs with the hatching eggs. This also should be done quietly at night when she is drowsy. The hatching eggs will be much cooler than the dummy eggs she has been sitting on, so try not to give her a nasty surprise. Slide a hatching egg under her and slip out a pot egg. If she accepts this, continue with the others until she is sitting on only the eggs you intend to hatch.

When selecting eggs for hatching, choose those produced by healthy hens who are consistently good layers. The hens and cockerel should be fully mature.

Select clean eggs of a standard size and shape, with no hairline cracks. If you need to mark them, use a soft pencil.

The eggs can be stored in a cool place (12 to 15°C/54 to 59°F) for about a week – the longer you keep them, the fewer will hatch. Store them pointed-end-downwards and alter their position daily (prop the egg-box at an angle and swap sides each day).

You can buy fertile eggs from breeders, poultry markets or even online (but make sure you are buying from a reputable source). Hatching eggs can be sent through the post, but courier is more reliable. Eggs that have travelled should rest for at least twelve hours in a cool place.

Eggs only start developing when incubation begins. They must all be put under the broody together so they hatch at the same time.

An odd number of eggs makes a better circle. A bantam may be able to cover about seven, while a large hen might manage eleven or more. If she has too many you will find eggs outside the nest. Remove these or a different one will end up on the outside each time the hen rearranges her eggs. This will result in a poor hatch.

Incubation

It's all up to the hen now. Put food and water where she can see it, but so she has to leave the nest in order to eat and drink. She will only do this once a day.

Feed the hen mixed corn, which is digested slowly and gives her the energy for her long fast. It also reduces the likelihood of her soiling the eggs.

Check she actually leaves the nest, as some hens refuse to budge once sitting. This isn't good for the hen or the eggs, which need to cool and absorb some oxygen. If necessary, gently lift her off the nest at the same time every day – check for any eggs caught under her wings or legs.

Keep the hen off the nest for about twenty minutes. Take the opportunity to discard any broken eggs, and use warm water to clean any that are badly soiled. Turn the eggs if necessary – a hen sitting too tightly may not be turning her eggs, which will result in the embryos becoming stuck to the shell membranes and dying. Make sure your hands are clean and try to handle the eggs as little as possible.

Check that the hen has done a dropping before she returns to the nest – this will be obvious as it will be very large and smelly!

Standard hens' eggs take twenty-one days to hatch and bantams' eggs about eighteen to twenty days. Other eggs vary – geese can take up to thirty-five days.

Candling

While the hen is off the nest you can 'candle' the eggs if you wish. This allows you to check which eggs are fertile and developing. Candling involves concentrating a bright light through the eggshell to view the contents (at one time a candle was used). You can buy an electric candler or make one from a small cardboard box with a light bulb or bright torch inside.

Cut a hole in the top of the box and sit the egg over it. You may need to gently turn the egg to see anything, which will be easier in a dark place (deep-brown eggs are the most difficult to candle). Don't continue for more than a few seconds, as the heat from the lamp could damage the embryo.

Wait until the egg has been incubating at least seven days before candling. A pattern of spidery veins should be visible – an infertile egg will be clear.

It's not essential to candle naturally hatched eggs, but it can be useful to know how many are fertile – infertile eggs should be discarded.

Hatching

Leave the hen undisturbed about three days before hatching is due – provide food and water, but don't worry if she doesn't leave the nest.

Before the eggs hatch, the hen will hear the chicks cheeping in their shells, and she will respond. It will be some time after this before the first chicks appear, and the whole process can take two or three days.

It will be tempting to peep in at the new family, but try not to disturb them. The chicks can survive for twenty-four hours on the remains of the yolk they take in immediately prior to hatching, but you could place some chick crumbs and water within their reach.

Use a water container the chicks can't fall into (a little plastic water tower or chick drinker). Wet chicks quickly become chilled, and they can drown in even a shallow dish.

The hen will come out of her broody trance when the eggs have hatched, immediately turning into an active and protective mother hen, ready to take on all dangers to defend her chicks.

Unhatched eggs

When she has decided that hatching is complete, the hen will abandon any remaining eggs. She usually knows best about this, and the eggs are unlikely to be viable. Shake them gently (be careful they don't burst!), and if you hear liquid inside, there is no chick.

You can also test eggs by putting them in a bowl of warm (hand-hot) water. If they move around, there are live chicks inside.

If a chick can't escape the egg it's likely to be a weak one – if you decide to help it out, it will need extra care in order to survive.

The new family

Soon after hatching, the hen will proudly lead her chicks into the run. She will teach them how to eat and drink – you only need to supply the materials.

Chicks need to eat regularly, so feed a generous ration of chick crumbs. The hen will eat some of these as well, calling the chicks to show them how it's done. She needs to restore her strength, so continue feeding corn, which she will tuck into enthusiastically – she may break some up for the chicks to eat too.

Provide some chick-sized insoluble (flint) grit to help with digestion. Never give young birds access to soluble grit or layers' feeds, as these will cause development problems.

Be prepared to replenish food and water containers regularly – the hen will now start scratching about energetically and making dust-baths.

While hen and chicks are happily occupied in the run, clean out the nest completely, replacing it with a thick layer of shavings.

The nest and run should be kept clean and dry. Don't put the drinker in the sleeping area where it will make the bedding damp. Raise it a little (not too much) on a paving slab or similar, to prevent the ground around becoming wet. Frequently move the coop to fresh ground, or renew the scratching material. Keep the chicks away from areas used by adult chickens, and make sure they are sheltered from prevailing weather. Remember that many creatures (including squirrels, magpies, rats and crows) like to eat little chicks.

If you have had a hatch during a particularly cold spell, you may need to provide a heat lamp (preferably one that doesn't produce light) at night.

The growing chicks

Gradually change from chick crumbs to growers' feed when the chicks are about six weeks old.

The hen usually stays with the chicks until they are around eight weeks old, then starts to lose interest. She may even peck at them. She can now return to the flock – watch out for any disputes. The chicks should stay in their run, although you may wish to separate the young cockerels.

Continue to monitor the youngsters throughout their development, and wait until they are fully-grown before introducing them to the adult birds. Ideally this should be just before the pullets start laying, so they can settle down in the flock without the added stress of producing their first eggs. They will be treated as strangers – follow the same procedure as you would when adding any new birds to an established flock (see Chapter 7).

Free-Range Chicks

When our chicks are around five weeks old we start allowing them into the garden for short periods. They have to be watched, but by giving them their 'playtime' in the afternoon, we can rely on Mum to get them back to their run (it's pretty impressive – one quick cluck and they all go to bed). Their time outside is gradually extended, and gives them a chance to learn from their mother the vital lessons they will need when free-ranging. Using this method, we have found that the hen eventually starts returning to the main henhouse at night, but re-joins the youngsters during the day, often staying with them for several months.

Using Artificial Incubation

With an incubator you can hatch eggs at any time, and don't have to wait until a broody hen is available. Artificial incubation isn't new – the ancient Egyptians were experts.

Although electric incubators have made the process much easier than it was in ancient times, skill and judgement still play an important part. Even experienced breeders have to accept occasional disappointments.

Artificial incubation is better attempted when you have some experience with chickens, as it not only involves hatching eggs but also raising chicks. For this reason, only a brief overview is included here. Whole books have been written about this subject, and if it is of interest it would be worth investing in a specialist guide.

Choosing an incubator

Incubators come in a range of sizes – from those that take half a dozen eggs to huge models that will hatch hundreds.

Apart from providing warmth, an incubator regulates humidity and airflow. The basic types require a fair amount of effort from the operator, while more sophisticated models will do most of the work, but are correspondingly more expensive.

With a manual incubator, the eggs have to be turned by hand a few times a day. Semi-automatic models allow you to turn the eggs from the outside, avoiding loss of heat and humidity. An automatic incubator takes care of turning the eggs, but will still require daily checks to ensure all is well.

Some incubators are 'still air', meaning there is no fan to keep the temperature standardized throughout, so you will need to check for cold spots. 'Forced-air' incubators have a fan to regulate temperature.

Choose an incubator from a reputable company, selecting one that is straightforward to use with clear instructions. It should be easy to maintain and clean – the incubator has to be thoroughly cleaned and disinfected before and after use.

Individual models differ, so follow the manufacturer's instructions carefully. Outside conditions will affect the operation of the unit, so install it somewhere dry, not too warm or cold, with no fluctuations of temperature. Run it for twenty-four hours and make sure it is functioning properly before adding the eggs. The recommended temperature for hatching is usually 37.5°C (99.5°F), but the setting may vary depending on the incubator.

Incubation

Choose eggs for hatching as suggested previously. They must be clean. Many people wash them first – use water slightly warmer than the eggs, adding an incubator disinfectant.

In the warm, moist environment of the incubator, good hygiene is essential. Wash your hands (or put on disposable gloves) before touching incubating eggs.

Candling allows you to discard infertile eggs, which may go bad and contaminate the others. Don't be tempted to add replacement eggs – the humidity will be wrong for them when the older ones reach the hatching stage, and they are likely to be infected by bacteria produced by the hatching chicks.

Adjust humidity according to the instructions for your incubator, but as the British climate tends to be damp anyway, be careful about adding too much water. The egg needs to lose moisture during incubation, and too-high humidity will result in the embryos drowning in their shells.

Humidity is usually increased a few days before hatching – if it's too low the shell membrane dries, making it impossible for the chick to break out.

Hatching

Stop turning the eggs two or three days before hatching is due.

On top of its beak the chick has an 'egg-tooth', which falls off after hatching. It uses this to make a small hole in the shell ('pipping'). It then slowly enlarges the hole and struggles out of the shell.

Don't open the incubator during hatching, as this will result in loss of humidity – helping chicks out of their shells can tear the attached blood vessels.

Leave the chicks for twenty-four hours to become dry and fluffy, then move them to the brooder.

The brooder

Chicks need a protected environment (the brooder) and a heat source. You can buy a purpose-made brooder or use a large, deep box. It must be draught-free, well-ventilated, and spacious enough to accommodate the chicks comfortably as they grow. Cover the top with mesh to prevent escapes and protect the chicks from inquisitive pets.

Find a quiet spot for the brooder, where the temperature remains fairly constant and doesn't become too hot. Chicks produce quite a bit of dust and mess, but if you decide to keep the brooder in an outbuilding, make sure it's completely safe from rats. The chicks will need light (either natural or artificial) during the day, but there should also be some hours of darkness at night.

Supplying heat

The heat source replaces the warmth provided by a broody hen. As in nature, the chicks will access the heat when necessary – but there must also be a cooler area in the brooder where they can eat, drink and exercise.

A traditional heat lamp consists of a powerful bulb with a metal shade, suspended from a hook and chain. The lamp is raised or lowered on the chain to decrease or increase the heat underneath. Avoid using a bright bulb, as the continuous light prevents the chicks resting at night, leading to stress-related problems. Infrared is better, and a ceramic bulb gives no light at all – but there is no obvious indication if it stops working. Loss of heat quickly causes casualties, so whichever bulb you use, always keep a spare handy.

The temperature under the lamp should be at 35°C (95°F) when the chicks are transferred from the incubator. It should then be reduced by two or three degrees each week until the chicks are fully-feathered and no longer need the extra warmth. Watch the chicks to see if the temperature is right – too hot and they will spread out away from the lamp; too cold and they'll huddle underneath it, cheeping loudly.

Heated bulbs (not surprisingly) get very hot. They must be securely fixed, and protected by a wire guard. Make sure the lamp isn't too close to flammable materials such as cardboard or bedding!

A more modern option is the electric panel. It stands on legs, and the chicks go underneath for warmth as they would with a hen. It costs more to buy than a heat lamp, but doesn't get nearly as hot, which makes it considerably cheaper to run as well as safer. Chicks are less likely to overheat, as only the area under the panel is warmed. With no hooks and chains to worry about, it's also easier and more convenient to use indoors.

Panels come in different sizes, based on the number of chicks being brooded, and the height is adjustable to accommodate the growing birds. Although panels have advantages over traditional heat lamps, there are a few points to bear in mind when deciding which to use:

- The chicks can't be easily checked when they are under the panel
- Droppings tend to stick to the panel, which requires frequent cleaning
- Care must be taken when positioning the panel in the brooder to avoid creating narrow gaps where chicks could become trapped
- In a very cold environment, the extra heat provided by a lamp may be a consideration

The heat source is usually the most expensive part of brooding chicks, so give some thought as to which type will best suit your personal circumstances.

Caring for the chicks

Watch the chicks closely, especially during the critical first few days. Provide plenty of chick crumbs, ideally medicated against coccidiosis unless the chicks have been vaccinated. Clean water is essential – use a drinker the chicks can't climb into, and keep both water and food away from the heat source. As the chicks have no mother hen to teach them, you may need to dip their beaks in the water, being careful it doesn't go into their nostrils. Scattering some chick crumbs on to a sheet of paper will encourage them to investigate, and they will soon learn to use the feeder. Supply some fine insoluble grit too.

Some people prefer not to use shavings in the brooder in case the chicks eat them. Having food constantly available will discourage them from eating their bedding, but you could use a slip-resistant mat covered with paper towels during the first few days. Newspaper is too slippery and can lead to leg deformities.

Keep bedding clean and dry (raise the drinker slightly or stand it on wire mesh). Remove droppings from the feeder and drinker, and clean them frequently. Dampness, poor hygiene and overcrowding will cause serious health problems. Hunched, listless chicks and diarrhoea are signs of coccidiosis (see Chapter 11), requiring prompt treatment. Immediately separate any chicks that appear unwell.

Chills and stress can lead to pasted vents, preventing the chicks from passing droppings. Use warm water to clean the vent area and dry carefully.

Growing chicks

Abrupt changes of temperature aren't good for young birds. Don't suddenly take away their heat lamp and move them to an outside pen. Huddling together for warmth at night may result in suffocation.

Depending on the time of year, the young birds can be moved to an indoor or outdoor rearing coop when they are fully feathered at about eight weeks. They need a warm, dry house, large enough for them all to shelter comfortably in bad weather. Feed can be gradually changed to growers' pellets; their ongoing care will be the same as for naturally hatched chicks.

Key Points

- Hybrids and light-laying breeds rarely go broody
- A broody hen has puffed-up feathers and her breast feels warm
- Hens shouldn't be allowed to remain broody if not required for hatching eggs
- Regularly collecting eggs helps discourage broodiness
- For a determined broody, a coop with a wire floor is the only remedy
- Watch out for 'secret broodies' in spring and summer
- Decide what will happen to the males before putting eggs to hatch
- Spring or early summer is the best time for natural hatching
- A separate broody coop is needed for hatching chicks
- Only hatch clean eggs of standard size and shape, with no cracks
- Let eggs rest in a cool place for twelve hours after travelling
- Put all the eggs under the broody at the same time
- Feed mixed corn to a sitting hen – she should leave the nest once a day
- Hens' eggs take twenty-one days to hatch, bantams' about eighteen to twenty days
- Consider vaccinations and protection against coccidiosis before the chicks hatch so that you are ready to act promptly
- Use a water container the chicks can't fall into, and feed chick crumbs
- Never give young birds access to layers' feeds or soluble grit
- Keep the chicks' environment clean and secure
- Don't introduce young birds to the flock until they are mature
- Artificial incubation requires some skill and experience
- An incubator provides warmth as well as regulating humidity and airflow
- Good hygiene is essential when using an incubator
- Candle the eggs, but don't replace infertile eggs with fresh ones
- Watch humidity levels carefully
- Stop turning eggs two or three days before hatching is due
- Avoid opening the incubator during hatching
- Helping chicks out of their shells can tear the attached blood vessels

- Artificially incubated chicks need a brooder with a heat source
- The chicks will indicate whether they are too hot or cold – check them regularly
- Artificially incubated chicks may not know how to eat and drink at first
- Dampness, poor hygiene and overcrowding will cause serious health problems
- Don't subject young birds to sudden changes of temperature

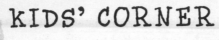

KIDS' CORNER

Quiz

Are you going to hatch some eggs? First, make sure you can answer these questions from Chapter 13.

QUESTION ONE

What should you do if a hen goes broody and you don't want her to hatch any chicks?

    (a) Leave her alone

    (b) Give her some pottery eggs to sit on

    (c) Discourage her from being broody

QUESTION TWO

When is the best time for a hen to start sitting on eggs?

    (a) Spring

    (b) Autumn

    (c) Winter

QUESTION THREE

Where is the best place for a hen to hatch her eggs?

    (a) In a separate broody coop

    (b) In the nest-box

    (c) Under a bush

QUESTION FOUR

When do eggs start to develop into chicks?

    (a) Before they are laid

    (b) As soon as they are laid

    (c) When the hen starts sitting on them

QUESTION FIVE

What is a brooder?

    (a) A machine for hatching eggs

    (b) A heated area where chicks live after hatching

    (c) A special feeder for young chicks

ANSWERS

One (c); Two (a); Three (a); Four (c); Five (b)

If you knew all the answers, you are well on the way to hatching your first chicks!

Chicken Chat

'Don't count your chickens before they are hatched': You never know how many eggs are going to hatch into chicks, and it's also true that you shouldn't rely on something happening until it actually happens! This proverb comes from Aesop's *Fables*.

Chicken Jokes

Why was the egg scared?
Because it was a little chicken!

What do you call a chicken in a shell-suit?
An egg!

What do you get if you cross a chicken with a duck?
A bird that lays down!

Something to Do . . .

Make a cheap and easy egg candler! You'll need a toilet roll tube, a small bright torch and some sticky tape.

Cut four slits in one end of the toilet roll tube and fit it around the torch handle, overlapping the slits. Secure with sticky tape.

Carefully place the egg on top of the tube and switch on the torch. What can you see? (This works best in a darkened room.)

CHAPTER FOURTEEN

Chickening Out: When Chickens Die

Chickens aren't hypochondriacs. They can hide illness remarkably well. A hen may go to bed looking absolutely fine, only to be found cold and stiff the next morning, leaving her bewildered owner wondering, 'What did I do wrong?'

Like any animals, chickens can die suddenly. A bad shock could cause a chicken to have a heart attack, or an inherent weakness may result in an untimely demise. 'Chickens just die; there's no point in getting upset about it,' I was told when pondering my first fatality.

Guilt is a natural reaction to loss, even if you are entirely blameless. Still, a little guilt can be useful if it prevents further casualties. Think about your chicken care routines. Is there anything that could have allowed a bird to become sick – such as a stressful situation or an attack of parasites?

Regularly checking your chickens will help identify potential problems – otherwise an unexpected death may be the first indication that something is wrong.

Losses suffered from predators are particularly distressing because this is often due to human error – forgetting to shut up the birds, ignoring that loose catch on the door, etc. All you can do is learn from the mistake, and take prompt action to secure any remaining chickens before the predator returns.

Calling it a Day

When a chicken is unwell, some difficult decisions may be necessary. If the flock is required to be cost-effective, a sick or unproductive bird will usually be despatched.

Chickens kept as family pets are usually treated more leniently – although not always as fairly. If a hen becomes ill, a swift end is better than allowing her to suffer in the hope she may recover.

The other option is veterinary treatment. Even so, there may come a point when cost and stress to the bird make further intervention undesirable. The vet will help with this decision, and euthanasia can be administered by injection.

Poultry keepers are always advised to learn to kill a chicken in case of emergency. If the thought makes you shudder, you should be prepared to phone an expert friend or call the vet.

Disposing of the Body

People often suggest eating a chicken that has died – it always surprises me. The meat could be contaminated by disease, and with our highly regulated meat industry it doesn't make sense to risk the family's health to save a few pounds at the butcher's.

If a chicken has been euthanized by the vet, it must not be eaten by either humans or animals. Birds that have been killed by predators are usually mangled, and infection may have been introduced from teeth and saliva.

So what should you do with the sad little bundle of feathers?

Technically you should dispose of it in accordance with the current DEFRA guidelines. All dead poultry (there is no distinction between pet and farm chickens) should be incinerated at a registered premises. DEFRA or your local council can provide details of these, or your local hunt may have an approved incinerator and be prepared to help for a small charge.

For the small-scale poultry keeper the vet is often the only viable legal option when disposing of dead chickens, and charges vary. Many people bury their dead birds but, apart from being contrary to the law, buried chickens may attract predators. Foxes can sniff out food as much as 45cm below ground. Finding the grave destroyed by a fox will be distressing – and now he knows where to come for dinner, he'll be back for more.

How Long Do Chickens Live?

If fortunate enough to avoid predators and illness, a chicken's lifespan may be around eight to ten years. This depends to a certain extent on the breed – some might go on longer, while others will be in decline after three or four years. Bantams are generally thought to be the longest lived, while pure-breeds often live longer than hybrids.

Well cared-for chickens have the best chance of a long life, but no matter how attentive you are, the time will come when the inevitable has to be faced.

Losing a Pet

Chickens are often kept as pets, and even those bought solely to provide eggs can quickly endear themselves to a family. It is sad to lose a bird you have known and tended, and some owners find it very distressing. 'Only a chicken' doesn't apply to a hen who had a name, sat on your lap and came into the kitchen for her breakfast. She had a character and was loved – her loss will be mourned.

For many children losing a pet is often their first experience of death, and they should be encouraged to talk about their feelings. Holding a small 'funeral' is often helpful; a celebration of life can ease the pain of loss.

After a Loss

If you need to know why a chicken has died, the vet can perform a post-mortem. This could be useful if you suspect a contagious disease, but might not be conclusive.

Make sure the other chickens are healthy and that any problems have been fully resolved before buying new birds, although if you are down to just one hen she should be given a companion sooner rather than later.

Losing chickens is part of keeping them, but there's nothing 'wrong' with feeling sad when it happens. Give your chickens the best possible life and, when the time comes, you will be able to offer a good home to some more lucky hens.

Key Points

- Chickens can die suddenly without any obvious reason
- Try to identify any mistakes to prevent further losses
- Culling is preferable to allowing a bird to suffer – otherwise the vet must be consulted
- The vet can administer euthanasia by injection
- Never eat a chicken that has died unless it was a healthy bird slaughtered for the table
- Dead chickens should be disposed of in accordance with current DEFRA guidelines
- Chickens may live about eight to ten years, depending on the breed
- Make sure the other chickens are healthy before buying replacement birds

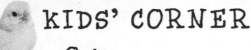

KIDS' CORNER

Quiz

It's sad to think about chickens dying, but you should know what to do when it happens. What did you learn from Chapter 14?

QUESTION ONE
Which of these statements is wrong?
    (a) Chickens are good at hiding illness
    (b) Chickens will soon let you know if they aren't well
    (c) Regular checks can help you to spot problems

QUESTION TWO
If a chicken is too ill to get better what should you do?
    (a) Leave her in peace to die with her friends around her
    (b) Have her put to sleep
    (c) Keep her warm and hope she will recover anyway

QUESTION THREE
What should you do with a chicken that has died of disease?
    (a) Eat it
    (b) Bury it in the garden
    (c) Have the body incinerated at a registered premises

QUESTION FOUR
What's the average lifespan of a chicken?
    (a) Two years
    (b) Eight to ten years
    (c) Twenty years

QUESTION FIVE
When chickens die, what should you do in the future?
    (a) Quickly buy some more
    (b) Stop keeping chickens in case you have done something wrong
    (c) Make sure there is no disease in the flock before introducing new hens

ANSWERS
One (b); Two (b); Three (c); Four (b); Five (c)

You now know all about chickens from eggs to eggsit!

Chicken Chat

'Walking on eggshells': You would have to step very lightly to walk on eggshells without breaking them! This saying describes handling a delicate subject or situation. For example: 'George was very sad when his favourite chicken died, and we tried not to upset him further – but talking to him was like walking on eggshells.'

Chicken Joke

What do you call a chicken ghost?
A poultry-geist!

Something to Do . . .

Start a chicken scrapbook so you will always remember the chickens you have kept. Give each hen a page (or more if you like). Write the name of the hen at the top of the page and add her photo or do a drawing. Include details of her breed, colour, age, the colour eggs she lays, any unusual characteristics, the date you bought her and anything else you feel is important.

You could also attach some of the hen's feathers or make a pattern out of pieces of her eggshells – wash feathers and eggshells with disinfectant before using them.

CHAPTER FIFTEEN

A Year in the Life:
Appreciating the Changing Seasons

If you have never much noticed the varying seasons before, keeping chickens will change your perspective. Each time of year brings something different.

Spring

This is the best time to start keeping chickens – or to hatch some chicks.

- Hens will start producing more eggs or come back into lay after winter
- Keep an eye on pullets coming into lay for the first time
- Make sure the hens have shelter from the elements
- Keep the run well maintained to avoid a mud-bath
- Repair any winter damage to the henhouse
- Parasites begin emerging in warm weather – check birds and housing regularly
- Don't forget internal parasites – this is a good time to administer Flubenvet
- Hens may go broody – decide whether to hatch some chicks this year
- Summer isn't far away, so start making holiday arrangements for your chickens

Summer

Relax in the garden with a chicken on your lap – and check her for parasites!

- Your chickens will be going to bed much later now
- Remember that foxes may be around during the long summer evenings

- Flightier breeds may fancy roosting in trees and should be discouraged
- If hens don't want to go into their house at night, check it for red mite
- Watch for hens making nests under bushes
- Consider freezing or selling surplus eggs
- Pick a sunny day to scrub the henhouse
- Be on the alert for parasites
- Deal with maintenance to the henhouse and run
- Make sure the chickens have shade – but remember they enjoy sunbathing too
- Place food and water out of the sun
- Keep the drinker full, but don't give icy water in hot weather
- Clean the drinker regularly to prevent algae forming
- Try to visit summer shows to see what's new in the chicken world
- If going out for the day, make sure you leave enough food and water
- Hens may start moulting at the end of summer

Autumn

Is the henhouse well-stocked? Can the chickens keep each other warm on chilly nights?

- Feed moulting chickens well to encourage re-feathering
- While the weather remains warm, keep checking for external parasites
- Treat with Flubenvet to deal with the summer crop of internal parasites
- Wash and disinfect the henhouse to kill off any bugs before winter
- Make sure the house and run are in good repair before the bad weather starts
- Prepare winter quarters if necessary – a large shed so the birds can stay indoors or a run on a solid surface
- Provide shelter from wind and rain
- Sudden storms might result in some odd eggs the next day
- Egg production may be declining
- The chickens will begin roosting earlier – make sure they are securely shut up at dusk
- The birds will eat more as the temperature drops, and they will appreciate some grain in the afternoon
- Rodents will be looking for comfortable winter quarters – don't attract them with food, and watch out for rat holes or gnawing

Winter

Winter nights are long and cold – keep your chickens dry and warm.

- Regularly check for leaks or draughts in the henhouse – make sure there is adequate ventilation too
- Provide plenty of dry bedding and change it regularly
- Be generous with feed – mixed corn given in the afternoon helps generate warmth overnight
- Supply some vegetables, but don't feed anything that's frosted
- Bring drinkers inside at night to prevent them freezing
- Check that water hasn't frozen during the day
- The chickens will be in bed by mid-afternoon – see that they are secure
- Snow can unsettle chickens – they may be reluctant to come out of their house
- Birds with feathered feet and legs find snow and wet particularly unpleasant
- Provide some shelter so the hens are encouraged to venture outside
- Predators and rodents are hungry now – ensure there are no weak spots in the house or run and watch for any evidence of their presence
- Apply Vaseline to large combs and wattles to help prevent frostbite
- Keep an eye open for external parasites if the weather turns mild
- Some hens may start laying again in January – proving that spring isn't that far away!

KIDS' CORNER

Quiz

You should know a lot about chickens by now. Try this last quiz to see what else you discovered in Chapter 15.

QUESTION ONE
Which of these statements is wrong?
- (a) Spring is a good time to start keeping chickens
- (b) Chickens should be treated for worms in the spring
- (c) Hens stop laying eggs in the spring

QUESTION TWO
Which of these tasks would you be likely to do in summer?
- (a) Refill the drinker more often
- (b) Give the chickens extra corn
- (c) Put Vaseline on large combs and wattles

QUESTION THREE
In autumn which of the following would you expect?
- (a) The hens to start laying more eggs
- (b) To find the chickens roosting earlier
- (c) Broody hens to be hatching their chicks

QUESTION FOUR
How would you help your chickens to stay warm in winter?
- (a) Feed some grain in the afternoon
- (b) Make sure they have lots of space to move around in their house
- (c) Close up all the ventilation holes in the coop

QUESTION FIVE
When should you check your chickens for external parasites?
- (a) In spring and summer
- (b) In autumn and winter
- (c) Whenever the weather is mild or warm

ANSWERS
One (c); Two (a); Three (b); Four (a); Five (c)

Well done if you knew all the answers – and good luck with your chicken keeping in the future!

Chicken Chat

'No spring chicken': A spring chicken is a young bird, and this saying is used about a woman who is no longer young. It often suggests that she is dressing or acting unsuitably. For example: 'Gran shouldn't be turning cartwheels at her age – after all, she's no spring chicken!'

Chicken Jokes

What happened to the hen who spent too much time sunbathing?
She laid fried eggs!

What do you call a chicken at the North Pole?
Lost!

Why did the chickens go out to play during the snowstorm?
Because it was fowl weather!

Something to Do . . .

Keep a chicken diary. Use a notebook or exercise book and write down your observations every day. Make a note of what each hen is doing, how she looks, whether she has laid an egg and if there are any worries about her health.

You can also use this book to keep a record of when you wormed your chickens, when they were given additives or supplements and if there have been any visits to the vet.

The longer you keep chickens, the more you will learn about them. Your chicken diaries will be useful reminders of all that you have discovered on your chicken-keeping journey.

Glossary

Air-sac Part of the breathing system – a balloon-like pocket of air

Air space Gap between the outer and inner membranes of the eggshell

Albumen Egg white

Anti-coccidiostat (ACS) Drug to help prevent coccidiosis

APHA Animal and Plant Health Agency

Ark Triangular henhouse, usually with attached run

As hatched Young chickens not yet sexed

Aspergillosis Respiratory disease caused by fungal spores

Auto-sexing breed Male and female chicks are different colours

Avian influenza Notifiable virus causing rapid death in birds, and potential infection in humans

Bantam Small chicken

Barred Two different-coloured stripes across each feather

Beard Clump of feathers under the beak

Blood spot Blood on the edge of the yolk

Breed standards Defined characteristics set down by the breed society

Breed true Parent birds produce young that resemble them

Brooder Heated area for raising chicks

Broody/broodiness Instinct to hatch eggs

Bumble foot Infected wound on the sole of the foot

Candling Shining a bright light through an egg to view its contents

Chalazae White 'strings' that hold the yolk in place

Chicken Generally used as the standard word for domestic fowl (originally chicken denoted a young bird)

Coccidia Parasite that causes coccidiosis

Coccidiosis Parasitic disease that destroys the gut walls

Cock Male after his first adult moult

Cockerel Male before his first adult moult – commonly used to describe all males

Comb Red fleshy appendage on the head

Convict's foot Wet litter and mud set firmly around the foot

Crest Tuft of feathers on the head

Crop Pouch at the base of the neck where food is stored after swallowing

Cuckoo Irregular barring where colours run together

Cuticle Protective covering of the eggshell

Cylinder feeder Central cylinder with surrounding trough

DEFRA Department for Environment, Food and Rural Affairs

Depluming mite Parasite that burrows into feather shafts

Diatomaceous earth Fossilized algae that damages the outer coating of insects

Droppings board Removable tray to catch droppings from roosting hens

Dual-purpose breeds Chickens that are both layers and meat birds

Dust-bath Dry soil for cleaning feathers and skin

Ear lobe Fold of skin beneath the ear

Egg-bound Hen unable to lay her egg

Egg peritonitis Yolks descend into the body cavity and cause infection

Egg-tooth Hard tip on a chick's beak that enables it to break out of the shell

Egg-withdrawal period Time during which eggs cannot be eaten following medication

Energizer Transforms power to run an electric fence

Enriched cage Improved accommodation for battery hens

Feather pecking Pulling out feathers

Flight feathers The ten long feathers at the end of each wing – also known as 'primaries'

Flighty Jumpy and easily startled

Flint/insoluble grit Small stones swallowed by chickens to enable digestion of food

Flubenvet Licensed chicken wormer

Forced air incubator Includes a fan to keep temperature constant throughout

Fowl pox Viral disease causing crusty scabs

Game breed Close-feathered birds, descendants of those originally bred for cock-fighting

Gapeworm Parasitic worm that attaches to the windpipe

Germinal disc/blastodisc Reproductive cell – a white spot on the edge of the yolk

Gizzard Muscular stomach where food is ground

Ground sanitizer Disinfectant that destroys worm eggs

Hard-feathered Tight feathering as seen on game fowl

Heavy breeds Weighty chickens that can be used as meat birds

Hen Female after her first adult moult

Horn comb Comb in the shape of a 'V'

Hybrid hen Layer developed by scientific cross-breeding
Impacted crop Blockage in the crop
Incubator Heated device for hatching eggs
Infectious bronchitis Respiratory disease that can damage the reproductive organs
Ivermectin Anti-parasitic medication not licensed for chickens
Laced Feathers edged with a different colour
Lavender Pale grey colour
Light breed Chicken primarily for eggs rather than meat
Marek's disease Virus causing paralysis, tumours and death
Mash Complete chicken feed in powdered form
Meat spot Small brown spot in the egg white
Millefleur Black and white patterning on a coloured background
Mixed corn Mixture of grains including maize
Moult Seasonal loss and replacement of feathers
Muff Extra feathers around the face
Mycoplasma Common and very infectious respiratory disease
Nest-box Compartment for egg-laying
Newcastle disease Notifiable virus causing many deaths in chickens
Northern fowl mite Parasite that lives on chickens causing serious debility
Notifiable disease An illness that must be reported to DEFRA
Oocyst Coccidia egg
Pair/trio/quartet/quintet Groups of chickens, all containing a cockerel
Pasted vent Blockage of droppings caused by chills and stress in chicks
Pea or triple comb Low knobbly comb with three ridges
Peck feeder Releases food when the chicken pecks at a spring
Pecking order Social structure of a flock
Pellets Small pieces of complete chicken feed
Pencilled Fine lines either around or across the feathers
Pin feather Small growing quill
Pipping When a hatching chick pierces the eggshell
Point of lay (POL) hen Female approaching the age when she is able to lay
Pop-hole Small door used by chickens to access the henhouse
Pot egg Imitation egg
Poultry saddle Device to protect the hen's back from the cockerel
Preen gland Organ producing oil for conditioning feathers
Preening Using the beak to tidy and oil the feathers
Prolapse When the oviduct protrudes outside the vent
Pullet Female before her first adult moult

Pure-breed Bird with defined characteristics, able to reproduce offspring that closely resemble the parents

Red mite Parasite that lives in the henhouse causing serious debility

Rollaway nest-box Laying compartment where eggs roll out of reach of the hens

Rooster American word for a male chicken

Roosting Sleeping on a perch or branch

Rose comb Flat, knobbly comb finishing in a spike

Salmonella Bacterial intestinal disease that can be passed between chickens and humans

Scaly leg mite Parasite that lives under the leg scales, causing irritation and lameness

Single comb Upright serrated comb

Soft-feathered Breeds other than game fowl

Soluble grit Usually crushed oyster shells to provide extra calcium

Sour crop Yeast infection of the crop

Spangled Contrasting colour at the end of each feather

Spurs Sharp horn-like growths on the legs of males and some hens

SQP Suitably qualified person to sell chemical wormers

Still air incubator Has no fan to regulate temperature

Strain Breeding line showing particular features

Tower drinker Central water reservoir with surrounding trough

Treadle feeder Releases food when the chicken stands on a metal plate

True bantam Small breed of chicken with no large equivalent

Utility strain Birds bred for productivity rather than showing

Vent Opening through which droppings and eggs are passed

Vetwrap Animal bandage that does not stick to hair or feathers

Waterglass Sodium silicate – used for preserving eggs

Wattles The flesh hanging under the beak

Wing clipping Trimming the flight feathers to prevent flying

Witch/fairy egg Tiny egg without yolk

Worm-egg count Laboratory test for worm-eggs in droppings

Further Reference

Organizations

Animal and Plant Health Agency
(Executive agency of DEFRA also working on behalf of the Scottish and Welsh governments, responsible for animal, plant and bee health)
www.gov.uk/government/organisations/animal-and-plant-health-agency

British Hen Welfare Trust
(Campaigns for better welfare for commercial layers and organizes re-homing of ex-battery hens)
Hope Chapel, Rose Ash, South Molton, Devon EX36 4RF
www.bhwt.org.uk
Tel: 01884 860084

Department for Environment, Food and Rural Affairs
Nobel House, 17 Smith Square, London SW1P 3JR
www.gov.uk/government/organisations/department-for-environment-food-rural-affairs
Tel: 03459 33 55 77

The Poultry Club of Great Britain
(Safeguards the interests of all pure and traditional breeds of poultry)
www.poultryclub.org

Breed Club Websites

Most of the breeds featured in Chapter 4 have a dedicated club with a website. Some may be found as groups on social media. The Poultry Club of Great Britain also has a list of breed clubs with contact details.

Autosexing Breeds Association
(i.e. Cream Legbar)
www.autosexing-poultry.co.uk/

Booted Bantam Society UK
www.bootedbantamsgroup.webs.com

British Araucana Club
www.thebritisharaucanaclub.co.uk

British Belgian Bantam Club
www.britishbelgianbantamclub.com

British Faverolles Society
www.faverolles.co.uk

Cochin Club
www.cochinclub.com

Dorking Club
www.poultryclub.org/dorkingclub/

Laced Wyandotte Club
www.thelacedwyandotteclub.co.uk

Leghorn Club
www.theleghornclub.com

Marans Club
www.themaransclub.co.uk

Orpington Club
www.theorpingtonclub.org.uk

Partridge and Pencilled Wyandotte Club
www.partridgewyandotteclub.co.uk

Pekin Bantam Club of Great Britain
www.pekinbantamclub.co.uk

Plymouth Rock Club
www.theplymouthrockclub.co.uk

Rare Poultry Society
(Covers breeds that don't have their
own club or society)
www.rarepoultrysociety.com

Rosecomb Bantam Club
www.rosecombbantamclub.co.uk

Sebright Club
www.sebrightclub.co.uk

Serama Club of Great Britain
www.seramaclubgb.co.uk

Silkie Club of Great Britain
www.thesilkieclub.weebly.com

Welsummer Club UK
www.welsummer.club

Magazines

These publications are either
exclusively about poultry or regularly
include poultry articles – many have
useful websites and forums.

Country Smallholding (Includes a
pull-out poultry magazine)
www.countrysmallholding.com

Fancy Fowl (Magazine for the
poultry fancier)
www.fancyfowl.com

Home Farmer (Regular features on
small-scale poultry keeping)
www.homefarmer.co.uk

Practical Poultry (For all levels of
poultry-keepers)
www.facebook.com/PracticalPoultry

Smallholder (Includes a section on
poultry)
www.smallholder.co.uk

Your Chickens (For the back-garden
hen keeper)
www.yourchickens.co.uk

Suppliers of Poultry, Equipment and Services

This is just a small selection of
suppliers – there are many others
and these are not intended as
personal recommendations.

Muirfield Hatchery (Sole breeder of
Black Rock hybrids)
R. & A. Lovett, Muirfield Hatchery,
Crosslee Poultry Farm, Bridge of
Weir, Renfrewshire PA11 3RQ
www.blackrockhens.co.uk
Tel: 01505 613075

Chicken School (Bespoke chicken-
keeping courses for small groups or
families, egg-hatching for schools
and consultancy service)
www.chickenschool.co.uk

Domestic Fowl Trust (Housing,
supplies, poultry and online shop)
Bell Brook Farm, Pigeon Green,
Snitterfield Lane, Snitterfield,
Stratford Upon Avon CV37 0LP
www.domesticfowltrust.co.uk
Tel: 01789 850046

Flyte so Fancy (Wide range of
housing, supplies, informative
website and online shop)
The Cottage, Pulham, Dorchester,
Dorset DT2 7DX
www.flytesofancy.co.uk
Tel: 01300 345229

Happy Hens (Breeder of Silkies and
Pekins based in North Herefordshire
– can arrange courier delivery)
www.happyhens.me.uk

W. & H. Marriage & Sons Ltd (Miller of animal feeds) Chelmer Mills, New Street, Chelmsford, Essex CM1 1PN
www.marriagefeeds.co.uk
Tel: 01245 612019

Mother Hen's Poultry (Several varieties of hybrids, some pure-breeds and other poultry, also equipment and feeds – based in Reading, Berkshire)
www.motherhenspoultry.com

Newland Poultry (Chickens, housing, supplies, chicken boarding service, informative website and online shop)
Newland Grange, Stocks Lane, Malvern, Worcestershire WR13 5AZ
www.newlandpoultry.com
Tel: 01684 216257

Omlet (Housing, supplies, informative website and online shop)
Tuthill Park, Wardington, Oxfordshire OX17 1RR
www.omlet.co.uk
Tel: 01295 750094

Rosie's Rare Breed Poultry (Pure-breed chickens, bantams, and housing – Devon/Cornwall border)
www.rosiesrarebreedpoultry.weebly.com
Tel: 01566 775246

Websites

You'll find plenty of information about chickens online, some of it extremely useful. However, there is also much that is merely personal opinion, inaccurate or relevant only to the country where the website is based. Here are some websites that have proved reliable.

Poultry Keeping and Veterinary Information

www.chickenvet.co.uk (Articles on health, diseases and general care; online shop; laboratory services, including worm-egg counts; list of chicken-friendly vets)

www.poultrykeeper.com (A vast resource of information on all aspects of poultry-keeping. There's something for everyone here, plus a forum)

www.vicvet.com (Website of poultry vet Victoria Roberts – information on health and care for all poultry; also advice service)

Miscellaneous

www.pegasusarchive.org/arnhem/pat_glover (The full story of Pat Glover and Myrtle, the parachuting chicken)

www.vwt.org.uk (Vincent Wildlife Trust, for details on stoats, weasels, polecats and pine martens)

www.wildlifeonline.me.uk (Extensive articles on foxes, with suggestions for deterring them from your garden and protecting poultry)

Acknowledgements

Although they probably won't read this, I must first acknowledge the help of numerous chickens. Over the years they have proved (fairly) patient instructors, and still enjoy testing me with the unexpected.

Next, my profound thanks and admiration to Jess Goodman, who created the delightful illustrations while dealing with the arrival of twins two months earlier than expected.

Thank you to Susannah and David Fielding for all their helpful suggestions, as well as for demonstrating how chickens can be kept in a small garden. Many thanks also to Anne Harding, who is always happy to share her years of experience with chickens, and provided an insight into managing a larger flock.

Numerous people have kindly shared their resources and photographs, but I would especially like to thank Mark Hickman from the Pegasus Archive, for allowing use of his extensive research into the story of Myrtle the parachuting chicken.

My immense gratitude to the editorial team at How To Books and Constable & Robinson for giving me the opportunity to write this book, as well as for their ongoing help and advice.

Finally, thank you to all my family and friends for their continuing encouragement. Most of all I appreciate the constant love and support of my husband, Martin, who delights in turning my dreams into reality.

Photo Credits

The author would like to thank the following for kindly providing photographs:

British Hen Welfare Trust – page 57(top left)
Flyte So Fancy – pages 19, 25(bottom), 27, 28, 29, 53 (top and middle), 54 (bottom), 55 (bottom), 59
Omlet – page 24 (top)
Hugo Wells – pages 24(bottom), 25(top), 57(bottom), 81
Wells Poultry Supplies – pages 53(bottom), 55(top), 56, 57(top right), 58
All other photos – author's collection

Index